Wissenschaftliche Reihe Fahrzeugtechnik Universität Stuttgart

Reihe herausgegeben von

André Casal Kulzer, Stuttgart, Deutschland

Hans-Christian Reuss, Stuttgart, Deutschland

Andreas Wagner, Stuttgart, Deutschland

Das Institut für Fahrzeugtechnik Stuttgart (IFS) an der Universität Stuttgart forscht interdisziplinär sowie technologieoffen an modernen und zukunftsorientierten Fahrzeugkonzepten. In enger Zusammenarbeit mit Partnern aus Industrie und Wissenschaft entstehen neue Lösungen für die Mobilität der Zukunft. Das Institut gliedert sich in drei spezialisierte Lehrstühle, die gemeinsam das gesamte Spektrum der Fahrzeugtechnik abdecken: Der **Lehrstuhl für Fahrzeugantriebssysteme** widmet sich der Forschung nachhaltiger Antriebslösungen für künftige Mobilitätskonzepte. Im Fokus stehen alternative, elektrische sowie hybride Antriebssysteme und deren Komponenten – einschließlich der Nutzung nachhaltiger Energieträger wie Wasserstoff, synthetischer Kraftstoffe und Batterien. Der **Lehrstuhl für Kraftfahrzeugmechatronik** beschäftigt sich mit vernetzten, intelligenten und adaptiven Fahrzeugen. Im Zentrum stehen Fragestellungen zum Automatisierten und Vernetzten Fahren, Diagnose, Ladetechnologien, verteilte Systeme sowie softwarebasierte Fahrzeugfunktionen. Der **Lehrstuhl für Kraftfahrwesen** erforscht die physikalischen Grundlagen der Auslegung zukünftiger Fahrzeugkonzepte. Im Mittelpunkt stehen die Bereiche Aerodynamik, Windkanaltechnik, Akustik/NVH, Fahrzeugdynamik, Reifenmanagement und Thermomanagement Gesamtfahrzeug. Das IFS verfügt über eine vielfältige und hochmoderne Forschungsinfrastruktur, die realitätsnahe Untersuchungen vom Einzelbauteil bis zum Gesamtfahrzeug ermöglicht. Besonders hervorzuheben sind der Multikonfigurations- und Antriebsprüfstand für komplexe Antriebskonzepte, der Stuttgarter Fahrsimulator zur Untersuchung menschlichen Fahrverhaltens, der Aeroakustik-Fahrzeugwindkanal für akustische und strömungstechnische Fragestellungen sowie der Thermowindkanal zur Analyse thermischer Prozesse im Gesamtfahrzeug. Die wissenschaftliche Reihe „Fahrzeugtechnik Universität Stuttgart“ dokumentiert die im Rahmen von Promotionen am IFS entstandene Beiträge zur Mobilität der Zukunft und zeigt deren thematische Vielfalt, methodische Tiefe und Praxisrelevanz.

Reihe herausgegeben von
Prof. Dr.-Ing. André Casal Kulzer
Lehrstuhl für Fahrzeugantriebssysteme
Institut für Fahrzeugtechnik Stuttgart
Universität Stuttgart
Stuttgart, Deutschland

Prof. Dr.-Ing. Andreas Wagner
Lehrstuhl für Kraftfahrwesen
Institut für Fahrzeugtechnik Stuttgart
Universität Stuttgart
Stuttgart, Deutschland

Prof. Dr.-Ing. Hans-Christian Reuss
Lehrstuhl für
Kraftfahrzeugmechatronik
Institut für Fahrzeugtechnik Stuttgart
Universität Stuttgart
Stuttgart, Deutschland

Dorsa Mohammad Zaheri

Safety-Oriented Optimization of Automotive E/E Architectures

Springer Vieweg

Dorsa Mohammad Zaheri
Institute of Automotive Engineering
(IFS), Chair in Automotive Mechatronics
University of Stuttgart
Stuttgart, Germany

Zugl.: Dissertation Universität Stuttgart, 2025
D93

ISSN 2567-0042 ISSN 2567-0352 (electronic)
Wissenschaftliche Reihe Fahrzeugtechnik Universität Stuttgart
ISBN 978-3-658-53044-0 ISBN 978-3-658-53045-7 (eBook)
https://doi.org/10.1007/978-3-658-53045-7

This Springer Vieweg imprint is published by the registered company Springer Fachmedien Wiesbaden GmbH, part of Springer Nature.
The registered company address is: Abraham-Lincoln-Str. 46, 65189 Wiesbaden, Germany

To my Parents & Hadi

Acknowledgements

I feel deeply fortunate to have crossed paths with so many remarkable individuals throughout my doctoral journey, whose support has been instrumental in the successful completion of this research and my academic progress. First and foremost, I would like to express my heartfelt gratitude to my advisor, Prof. Dr.-Ing. Hans-Christian Reuss, for his constant encouragement, patience, and invaluable guidance throughout my PhD. I am profoundly thankful for the trust he placed in me and for the opportunity to pursue my doctoral studies within his research group.

I also extend my sincere thanks to my dissertation committee members, Prof. Dr.-Ing. Michael M. Resch, and Prof. Dr.-Ing. Marco Huber for their insightful contributions and advice during the preparation and review of this work. Their feedback has been invaluable to my research. Furthermore, I am deeply grateful to Dr. Dan Greiner, my project manager, for his understanding and for providing me the time to focus on my thesis, even amidst the challenges posed by other ongoing projects.

My deepest thanks go to my husband, Hadi, whose patience, love, and unwavering support have been a source of strength throughout the past five years. This journey would not have been possible without the sacrifices made by my parents, Mary and Mahmoud, and my siblings, Damoon and Dila. Their absence during these years has been deeply felt, and I have cherished every memory of their presence.

I would also like to thank my friends and colleagues at IFS and FKFS for their continuous support and encouragement throughout this journey. In particular, I would like to sincerely thank Dennis Niedballa and Jonas Maier, with whom I collaborated closely throughout my PhD. Our many discussions over the past few years have profoundly shaped and enriched this work.

To everyone who has touched my life and contributed to this journey, I offer my deepest gratitude. Your presence and support have been a guiding light, and I am forever thankful for each of you.

Dorsa Zaheri

Contents

List of Figures

List of Tables

List of Algorithms

List of Abbreviations

AAOL	Automotive architecture optimization language
ACO	Ant colony optimization
ACP	Automotive cyber-physical system
ADAS	Advanced driver assistance system
AHP	Analytical hierarchy process
ASIL	Automotive safety integrity level
CAN	Controller area network
CCF	Common cause failure
CSP	Constraint satisfaction problem
DAG	Directed acyclic graph
DAL	Design assurance levels
DSE	Design space exploration
E/E	Electrical/Electronic
EA	Evolutionary algorithm
ECU	Electronic control unit
EMF	Eclipse modeling framework
EV	Electric vehicle
FFI	Freedom from interference
FMEA	Failure modes and effects analysis
FMEDA	Failure modes, effects, and diagnostic analysis
FSC	Functional safety concept
FSR	Functional safety requirement
FTA	Fault tree analysis
GA	Genetic algorithm
HARA	Hazard analysis and risk assessment

HPC	High-performance computing
ILP	Integer-linear programming
ISO	International organization for standardization
JRaS	Joint routing and scheduling
LFM	Latent fault metric
LIN	Local interconnect network
LKS	Lane-keeping system
LQN	Layered queueing network
MBSE	Model-based systems engineering
MCS	Minimum cutsets
MILP	Mixed-integer linear programming
MIP	Mixed-integer programming
MOEA	Multi-objective evolutionary algorithm
NLP	Non-linear programming
NP	Non-deterministic polynomial time
OEM	Original equipment manufacturer
OTA	Over-the-air
PB	Pseudo-Boolean
PMHF	Probabilistic metric for random hardware failure
POF	Probability of failure
PSO	Particle swarm optimization
QM	Quality management
SA	Simulated annealing
SDV	Software defined vehicle
SG	Safety goal
SIL	Safety integrity levels
SMT	Satisfiability modulo theories

SOA	Service-oriented software architecture
SOTIF	Safety of the intended functionality
SPFM	Single point fault metric
SysML	Systems modeling language
TSN	Time-sensitive networking
TSR	Technical safety requirement
TT	Time-triggered
UML	Unified modeling language
WCET	Worst-case execution time
WCRT	Worst-case response time

Notations

	Applications
A	Set of all applications
$A_p \in A$	An application
D	Set of all edges in application graph
$G(T, D)$	Application graph
$T = \{T_1, \ldots, T_n\}$	Set of all tasks of applications
T'	Set of tasks that require ASIL decomposition $T' \subseteq T$
$T_{i,d}$	Deadline of task T_i
$T_{i,p}$	Period of task T_i
Λ_{T_i}	ASIL level of $T_i \in T$
$d_{i,j} \in D$	Dependency between T_i and T_j
$mem_{T_i,h}$	Memory of T_i with ASIL h
$wcet_{i,k,h}$	WCET of T_i running on N_k with ASIL h

	Communication messages
$M = \{M_1, \ldots, M_n\}$	Set of all communication messages
${M_m}^{dst}$	Destination task of communication message M_m
${M_m}^{src}$	Source task of communication message M_m
$M_{m,d}$	Deadline of the communication message M_m
$M_{m,p}$	Period of the communication message M_m
$M_{m,s}$	Size of the communication message M_m

	Decision variables
$\theta, \gamma, \delta, \eta$	Binary: support variables
$\alpha_{i,h}$	Integer: number of redundant tasks with ASIL h for T_i
β_k	Integer: maximum ASIL level assigned to E_k

$\sigma_{m,k}$	Continuous: release time of message M_m on node N_k
$\tau_{i,k}$	Continuous: starting time of T_i on N_k
$x_{i,k,h}$	Binary: 1 if task T_i with ASIL h executed on N_k
$y_{m,k,n,r,q}$	Binary: 1 if message M_m sent from N_k to N_n over $l_{r,q}$

	Hardware components
$E = \{E_1, \ldots, E_m\}$	Set of all processing elements
$N = \{N_1, \ldots, N_n\}$	Set of all vehicle nodes, $N = E \cup S$
$S = \{S_1, \ldots, S_l\}$	Set of all network switches and gateways
Λ_{N_k}	ASIL level of $N_k \in N$
λ_k	Failure rate of N_k
$br_{r,q}$	Bus speed between nodes N_r and N_q
$loc_{i,k}$	Feasibility of running T_i on N_k
$l_{r,q}$	Physical path between nodes N_r and N_q
mem_{N_k}	Memory of N_k
$rl_{r,q}$	Number of paths between nodes N_r and N_q
$wcrt_{i,j}$	WCRT of a communication message sent from T_i to T_j

Kurzfassung

Die Automobilindustrie befindet sich derzeit in einem tiefgreifenden technologischen Wandel. Getrieben wird dieser Wandel durch die fortschreitende Digitalisierung sowie durch stetig steigende Anforderungen an funktionale Sicherheit, Konnektivität, Energieeffizienz und Automatisierung. Fortschrittliche Fahrerassistenzsysteme (ADAS), hochautomatisierte und autonome Fahrfunktionen, die zunehmende Vernetzung von Fahrzeugen sowie die Elektrifizierung des Antriebsstrangs verändern die Entwicklung moderner Fahrzeuge grundlegend. Während Fahrzeuge in der Vergangenheit überwiegend durch ihre mechanischen und elektrischen Komponenten charakterisiert wurden, werden heutige und zukünftige Fahrzeuggenerationen zunehmend durch ihre Software bestimmt. Diese Entwicklung führt zum Paradigma des softwaredefinierten Fahrzeugs (SDV), bei dem Fahrzeugfunktionen nicht mehr ausschließlich durch dedizierte Hardware implementiert werden, sondern durch Software, die unabhängig von der zugrunde liegenden Hardware entwickelt, bereitgestellt, aktualisiert und erweitert werden kann. Dadurch entsteht ein langfristig evolvierbares Systemkonzept, das kontinuierliche Funktionsverbesserungen, neue Dienste sowie iterative Softwareupdates über den gesamten Lebenszyklus eines Fahrzeugs ermöglicht und klassische, hardwarezentrierte Entwicklungszyklen grundlegend transformiert.

Mit dem Übergang zum softwaredefinierten Fahrzeug verändern sich auch die Anforderungen an die zugrunde liegenden Elektro-/Elektronik-Architekturen (E/E-Architekturen) grundlegend. Traditionelle verteilte Architekturen mit einer großen Anzahl spezialisierter elektronischer Steuergeräte (ECUs) stoßen hinsichtlich Skalierbarkeit, Wartbarkeit, Kosten und Integrationsfähigkeit zunehmend an ihre Grenzen. Diese stark fragmentierte Architektur führt zu komplexen Kommunikationsbeziehungen, hoher Verkabelungsdichte sowie ineffizienter Ressourcennutzung und erschwert gleichzeitig die Einführung neuer Fahrzeugfunktionen. Moderne Fahrzeugplattformen verfolgen daher den Übergang zu zentralisierten beziehungsweise zonalen E/E-Architekturen, in denen eine deutlich reduzierte Anzahl leistungsstarker Recheneinheiten, insbesondere High Performance Computers (HPCs), die Funktionalität vieler klassischer Steuergeräte konsolidiert übernimmt. Parallel dazu ermöglichen serviceorientierte Softwarearchitekturen die Entwicklung modularer, hardware-unabhängiger Softwarekomponenten, die flexibel eingesetzt, aktualisiert und über den gesamten Lebenszyklus hinweg erweitert werden können. Diese Entkopplung von Hard- und Software bietet erhebliche Vorteile hinsichtlich Flexibilität, Skalierbarkeit und Wiederverwendbarkeit, führt jedoch gleichzeitig zu einer signifikanten Erhöhung der Systemkomplexität, insbesondere in Bezug auf Ressourcenverwaltung, Interferenzkontrolle, Timing-Vorhersagbarkeit und systemweite Integrationsvalidierung.

Mit der steigenden Anzahl softwarebasierter Funktionen wächst auch die Bedeutung der funktionalen Sicherheit in modernen Fahrzeugarchitekturen erheblich. Zahlreiche Fahrzeugfunktionen besitzen sicherheitskritischen Charakter und müssen den Anforderungen der Norm ISO 26262 entsprechen. Diese Norm definiert einen strukturierten, risikobasierten Entwicklungsprozess für sicherheitsrelevante elektrische und elektronische Systeme und umfasst Methoden zur Gefahrenanalyse, Risikobewertung, Systemgestaltung sowie Verifikation. Zentrales Element ist die Klassifikation von Funktionen anhand des Automotive Safety Integrity Level (ASIL), welcher die erforderlichen Sicherheitsmaßnahmen und Entwicklungsprozesse bestimmt. Dabei stellt ASIL-D die höchste Kritikalitätsstufe dar und erfordert besonders strenge Nachweise hinsichtlich Fehlervermeidung, Fehlererkennung und Fehlerbeherrschung. Ein wesentliches quantitatives Maß innerhalb der Norm ist das Probabilistic Metric for Hardware Failure (PMHF), welches die Wahrscheinlichkeit gefährlicher Hardwareausfälle beschreibt und als zentrale Bewertungsgröße für sicherheitskritische Systemarchitekturen dient. Da moderne Hochleistungsrechner sowohl sicherheitskritische als auch nicht sicherheitskritische Funktionen gleichzeitig ausführen, entstehen zusätzliche Herausforderungen hinsichtlich deterministischer Systemverhalten, zeitlicher Isolation, räumlicher Trennung sowie der Gewährleistung von Freedom from Interference (FFI). Diese Anforderungen müssen bereits in frühen Entwurfsphasen berücksichtigt werden, da sie tief in Architekturentscheidungen eingreifen und die spätere Systemzertifizierung maßgeblich beeinflussen.

Die effiziente Zuordnung von Softwarefunktionen auf verfügbare Hardwareressourcen stellt daher eine der zentralen Herausforderungen zukünftiger Fahrzeugarchitekturen dar. Neben funktionalen Sicherheitsanforderungen müssen zahlreiche weitere Randbedingungen berücksichtigt werden, darunter Rechenkapazitäten, Energieverbrauch, Kommunikationslatenzen, Zuverlässigkeitsanforderungen, Scheduling-Bedingungen sowie wirtschaftliche Aspekte wie Entwicklungs-, Integrations- und Hardwarekosten. Diese Zielgrößen stehen häufig in komplexen Zielkonflikten zueinander, wodurch ein hochdimensionales kombinatorisches Optimierungsproblem entsteht. Die besondere Schwierigkeit ergibt sich dabei aus der engen Kopplung zwischen Funktionsallokation, zeitlicher Ausführung, Kommunikationsstruktur und Sicherheitsrestriktionen. Bereits kleine Änderungen in der Architekturzuweisung können erhebliche Auswirkungen auf das gesamte Systemverhalten haben, insbesondere auf Latenzen, Netzwerkauslastung und sicherheitsbezogene Eigenschaften.

Vor diesem Hintergrund wird in dieser Dissertation ein generisches, modellbasiertes Framework zur automatisierten Exploration des Entwurfsraums moderner automobiler E/E-Architekturen vorgestellt. Ziel des Frameworks ist die automatisierte Synthese optimierter Systemarchitekturen unter gleichzeitiger Berücksichtigung funktionaler Sicherheitsanforderungen, Echtzeitbedingungen und wirtschaftlicher Randbedingungen. Das Framework ermöglicht die systematische Zuordnung von Softwarefunktionen zu Hardwareplattformen, die Optimierung von Kommunikationsstrukturen sowie die Einhaltung sicherheitsrelevanter Systembedingungen. Dadurch wird der manuelle Entwurfs- und Integrationsaufwand erheblich reduziert und gleichzeitig eine strukturierte und reproduzierbare Exploration großer Designräume ermöglicht, in denen komplexe Architekturentscheidungen quantitativ bewertet und systematisch verglichen werden können.

Die Hardware wird dabei als verteiltes heterogenes System modelliert, bestehend aus Hochleistungsrechnern und weiteren Recheneinheiten, die über Kommunikationsnetzwerke wie Automotive Ethernet verbunden sind. Die Software wird durch einen gerichteten azyklischen Graphen (DAG) beschrieben, in dem Knoten Funktionen oder Tasks und Kanten deren Datenabhängigkeiten darstellen. Auf dieser Grundlage wird ein Mixed-Integer-Programming-(MIP)-

Modell entwickelt, das die gleichzeitige Optimierung von Aufgabenallokation, Scheduling und Kommunikationsplanung ermöglicht.

Ein wesentlicher wissenschaftlicher Beitrag dieser Dissertation besteht in der integrierten Optimierung mehrerer bislang überwiegend getrennt betrachteter Entwurfsprobleme. Während bestehende Ansätze typischerweise Funktionsmapping, Task-Scheduling oder Kommunikationsplanung isoliert behandeln, integriert das vorgeschlagene Framework diese Aspekte gemeinsam mit sicherheitsrelevanten Anforderungen gemäß ISO 26262 in einem einheitlichen Mixed-Integer-Programming-(MIP)-Modell. Dadurch wird eine konsistente gemeinsame Optimierung von Funktionsausführung, zeitlicher Planung und Kommunikationsverhalten ermöglicht. Insbesondere können komplexe Interdependenzen zwischen Rechenlast, Netzwerklast und sicherheitskritischen Anforderungen explizit modelliert werden, wodurch global konsistentere und effizientere Systemlösungen entstehen.

Ein zentrales Merkmal des vorgeschlagenen Ansatzes ist die systematische Überführung der Anforderungen der ISO 26262 in ein mathematisches Optimierungsmodell. Sicherheitsanforderungen werden dabei nicht nachträglich überprüft, sondern bereits während der Architektursynthese als formale Nebenbedingungen integriert. Neben strukturellen und zeitlichen Sicherheitsrestriktionen werden insbesondere probabilistische Sicherheitsmetriken wie das PMHF direkt in die Optimierung eingebunden. Darüber hinaus werden Anforderungen an Fehlerisolation, deterministische Ausführung, räumliche Trennung sowie zeitliche Vorhersagbarkeit berücksichtigt. Dadurch entsteht ein sicherheitsgetriebener Entwurfsprozess, in dem zulässige Architekturen systematisch generiert werden und die Einhaltung normativer Anforderungen bereits inhärent gewährleistet ist.

Ein weiterer wesentlicher Beitrag der Arbeit ist die explizite Modellierung der ASIL-Dekomposition als Entscheidungsvariable innerhalb des Optimierungsproblems. Die ASIL-Dekomposition ermöglicht die Zerlegung sicherheitskritischer Funktionen in redundante oder teilweise unabhängige Teilfunktionen, um Sicherheitsanforderungen effizienter zu erfüllen und gleichzeitig Ressourcen effizienter zu nutzen. Gleichzeitig führt diese Zerlegung jedoch zu zusätzlicher Kommunikationslast, erhöhter Scheduling-Komplexität sowie zusätzlichem Rechenaufwand. Das vorgeschlagene Modell berücksichtigt diese Wechselwirkungen explizit und ermöglicht eine automatische Abwägung zwischen Sicherheitsgewinn, Kostenreduktion und zusätzlicher Systemkomplexität.

Dabei werden zwei grundlegende Entwurfsstrategien gemäß ISO 26262 berücksichtigt. Im Mixed-ASIL-Ansatz können Funktionen unterschiedlicher Sicherheitsstufen auf derselben Hardwareplattform koexistieren, sofern geeignete Maßnahmen zur Gewährleistung von Freedom from Interference (FFI) implementiert sind. Diese umfassen insbesondere zeitliche Partitionierung, räumliche Isolation sowie kontrollierte Ressourcennutzung innerhalb gemeinsam genutzter Rechensysteme. Im Max-ASIL-Ansatz hingegen führt die gemeinsame Ausführung unterschiedlicher ASIL-Stufen dazu, dass die gesamte Hardwareplattform auf das höchste Sicherheitsniveau angehoben werden muss, was erhebliche Auswirkungen auf Kostenstruktur, Hardwareanforderungen und Zertifizierungsaufwand hat. Beide Strategien werden innerhalb des MIP-Modells formal beschrieben und systematisch vergleichbar gemacht, wodurch eine quantitative Bewertung ihrer Auswirkungen möglich wird.

Neben der Funktionszuordnung und Sicherheitsmodellierung integriert das Framework auch die Kommunikationsplanung in den Optimierungsprozess. Insbesondere nach einer ASIL-Dekomposition entstehen zusätzliche Kommunikationsbeziehungen, da redundante oder aufgeteilte Funktionen Daten zwischen verteilten Komponenten austauschen müssen, um Konsistenz und funktionale Korrektheit sicherzustellen. Das Modell berücksichtigt daher gleichzeitig Routing- und Scheduling-Entscheidungen im Kommunikationsnetzwerk und koppelt diese direkt an die Funktionsallokation. Unterstützt werden verschiedene Routingstrategien, darunter Single-Path Routing, FFI-basiertes Routing mit Pfadisolation sowie Disjoint-Path Routing mit vollständiger Trennung von Netzwerkressourcen. Diese Strategien ermöglichen die flexible Abbildung unterschiedlicher Anforderungen an Latenz, Zuverlässigkeit und Sicherheitsisolation in Ethernet-basierten Fahrzeugnetzwerken und erlauben eine systematische Optimierung der Kommunikationsarchitektur.

Die Leistungsfähigkeit und praktische Anwendbarkeit des vorgeschlagenen Frameworks werden anhand umfangreicher Fallstudien auf realitätsnahen automobilen E/E-Architekturen validiert. Dabei werden unterschiedliche Systemgrößen, Kommunikationslasten und Sicherheitskonfigurationen untersucht, um die Skalierbarkeit und Robustheit des Ansatzes unter realistischen Bedingungen zu bewerten. Zusätzlich wird ein Kostenmodell für die Softwareentwicklung in den Optimierungsprozess integriert, wodurch sowohl technische als auch wirtschaftliche Auswirkungen von Architekturentscheidungen systematisch analysiert werden können. Ergänzend werden geeignete Bewertungsmetriken eingesetzt, um die Qualität und Effizienz der erzeugten Lösungen umfassend zu charakterisieren und miteinander vergleichbar zu machen.

Die experimentellen Ergebnisse zeigen, dass das Framework in der Lage ist, qualitativ hochwertige E/E-Architekturen automatisch zu erzeugen und dabei den manuellen Entwicklungsaufwand signifikant zu reduzieren. Darüber hinaus führt die integrierte Optimierung von Funktionsmapping, Scheduling, Kommunikationsdesign und Sicherheitsanforderungen zu einer deutlich konsistenteren Gesamtsystemarchitektur im Vergleich zu Ansätzen, die diese Teilprobleme isoliert betrachten. Insbesondere wird deutlich, dass die gemeinsame Modellierung aller relevanten Entwurfsdimensionen entscheidend für die Qualität und Effizienz der resultierenden Systemarchitektur ist.

Der Vergleich mit etablierten Optimierungsverfahren zur Ressourcenallokation zeigt zudem, dass der vorgeschlagene MIP-basierte Ansatz insbesondere bei mittelgroßen Problemstellungen robuste und qualitativ hochwertige Lösungen liefert und gleichzeitig eine hohe Modellierungsflexibilität aufweist. Zur weiteren Verbesserung der Rechenzeit werden zusätzliche Strategien wie binäre Suchverfahren zur Einschränkung des Lösungsraums sowie eine priorisierte Behandlung einzelner Zielfunktionen eingesetzt, wodurch eine effizientere Exploration des komplexen Designraums ermöglicht wird. Die Ergebnisse unterstreichen die Fähigkeit des Ansatzes, den Entwurfsprozess von E/E-Architekturen signifikant zu strukturieren und zu beschleunigen sowie gleichzeitig robuste und qualitativ hochwertige Lösungen für zukünftige Fahrzeuggenerationen bereitzustellen.

Abstract

The automotive industry is undergoing significant structural transformations, driven by technological advancements such as advanced driver assistance systems (ADAS), enhanced connectivity, and autonomous driving capabilities. These innovations have introduced new functionalities and increased the complexity of vehicle electronic systems, leading to the emergence of software-defined vehicles (SDVs). The concept of SDVs represents a major trend in automotive design, relying on the separation of hardware and software to enable easier updates and maintenance.

To support these advancements, automakers are adopting centralized computing and zonal electrical/electronic (E/E) architectures. Next-generation vehicles increasingly incorporate advanced in-vehicle communication protocols and a few high-performance computing units to manage both non-safety and safety-critical functionalities. As these systems evolve, ensuring functional safety has become a priority, requiring adherence to standards such as ISO 26262 to mitigate risks in safety-critical E/E systems during the design phase. Consequently, the seamless integration of software into hardware is essential for ensuring system quality. However, the growing complexity of the design process makes manual resource allocation not only time-consuming but also prone to suboptimal outcomes.

To address these challenges, this thesis proposes a generic framework designed to automatically explore E/E architectures and generate optimized solutions. It deploys software onto hardware components while satisfying both safety and non-safety requirements using a mixed-integer programming (MIP) model.

Hardware is modeled as a distributed system consisting of heterogeneous computing units connected via Ethernet or other communication networks, while the software is represented using a directed acyclic graph (DAG) model. The algorithm is capable of solving multi-objective optimization problems by prioritizing objectives in descending order, making it adaptable to a variety of applications and offering a versatile tool for E/E architecture design.

Compared to existing frameworks, the proposed approach includes an expanded range of safety analyses compliant with the ISO 26262 standard, addressing reliability, task scheduling, and automotive safety integrity level (ASIL) compatibility. A key feature of the framework is automatic ASIL decomposition, which identifies cost-efficient solutions, particularly when hardware resources for safety-critical applications are limited. The framework also evaluates the impact of two ISO 26262 design strategies: Mixed-ASIL design, in which software components of different safety levels—including both safety-critical and non-safety-critical components—coexist on the same hardware, and Max-ASIL design, in which all software is developed according to the highest ASIL level among the components deployed on that hardware.

The optimal message routing and scheduling are also integrated into the algorithm, facilitating efficient network topology design. Additional constraints, such as freedom from interference (FFI) and the independence requirements specified by ISO 26262 for ASIL decomposition, can be incorporated to ensure a safe and robust architecture.

The proposed approach is validated through a comprehensive case study, demonstrating its effectiveness in practice. Furthermore, this thesis compares the performance of the proposed optimization algorithm with state-of-the-art algorithms for resource allocation, highlighting both the advantages and limitations of the proposed method. The potential application of techniques such as binary search heuristics to reduce algorithm execution time is also explored, offering avenues for enhancing computational efficiency. Extensive experiments further evaluate the framework's scalability and practicality, underscoring its capability to streamline the E/E architecture design process and deliver robust solutions for next-generation automotive systems.

1 Introduction

Automotive electrical/electronic (E/E) architectures serve as the fundamental structure of the entire vehicle system, encompassing all E/E functions and components. They define the relationships among in-vehicle components, their interactions with the external environment, and enable the transmission of electrical and electronic signals. These architectures play a crucial role in shaping the functionality and performance of automotive systems, highlighting their significance in the development process [1].

In recent years, the integration of an increasing number of sensors within and around vehicles to enhance safety has significantly raised the complexity of E/E architectures. This trend has been driven largely by the adoption of advanced technologies such as advanced driver assistance systems (ADAS), autonomous driving, electrification, connectivity, and intelligent systems. As a result, both the number of E/E components and the volume of software embedded in vehicles have grown rapidly [2, 3]. Modern vehicles now often incorporate dozens to over a hundred electronic control units (ECUs), running up to 100 million lines of software code. This surge in complexity presents significant challenges in the design, development, validation, and maintenance of automotive E/E systems [4].

The increasing reliance on sophisticated software demands not only more powerful hardware but also makes the integration of such software into hardware more difficult [5]. Additionally, E/E architectures must comply with stringent development standards, such as ISO 26262, which ensures functional safety in passenger vehicles. Addressing safety requirements early in the development lifecycle is critical to preventing design flaws that could lead to accidents or other safety issues [6].

Future E/E architectures need to evolve to tackle these challenges by reducing the cost and complexity of wiring harnesses while also meeting safety requirements. However, the design process is becoming more complicated due to the increasing complexity of software, hardware, and their integration. These

D. Mohammad Zaheri, *Safety-Oriented Optimization of Automotive E/E Architectures*, Wissenschaftliche Reihe Fahrzeugtechnik Universität Stuttgart,
https://doi.org/10.1007/978-3-658-53045-7_1

elements must still be considered together to ensure that the systems achieve optimal performance and meet safety standards [5].

1.1 Trends in Automotive E/E Architectures

The early 2020s have witnessed a rapid shift toward intelligent electric vehicles (EVs), marking the beginning of a transformative era in the automotive industry. To cope with current and future needs of the industry, fundamental changes in E/E architecture are required. As a critical enabler, E/E architecture influences key factors such as cost, weight, and sustainability. Therefore, its evolution has garnered considerable attention from both industry and academia [7], [8]. Traditional E/E architectures, which rely on decentralized, mechanically-oriented designs with individual electronic control units (ECUs) handling specific functions, are no longer scalable enough to accommodate the growing complexity of modern vehicle systems. To overcome these limitations, the industry is transitioning toward a software-oriented design paradigm, enabling enhanced in-vehicle communication and expanded functionality. Figure 1.1 demonstrates the history and envisioned future of E/E architectures. As it is shown, modern E/E architectures are expecting to evolve from decentralized systems to cross-domain architectures and, ultimately, to centralized computing platforms [9, 10].

Although decentralized E/E architectures offer advantages such as simpler ECU integration, lightweight components, and the use of average processing power, they face limitations in handling the increasing complexity of modern vehicles. The "one function per ECU" approach leads to higher costs, wiring challenges, and restricted design flexibility. Traditional networks like controller area network (CAN) and local interconnect network (LIN) are also becoming insufficient, prompting a shift toward automotive Ethernet, which supports higher data volumes and time-critical communication.

To overcome these challenges, the next expected stage is domain-centralized architecture. In this approach, ECUs are grouped based on functionality, dividing vehicles into distinct domains such as infotainment, powertrain, body,

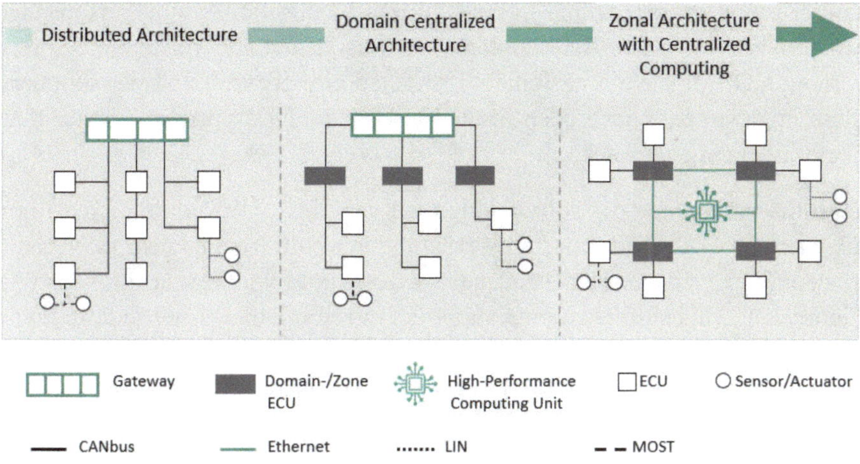

Figure 1.1: Evolution of vehicular E/E architectures [10]

and ADAS. However, domain-centralized architectures fall short in meeting the demands of modern vehicles due to the complexity of domain-specific modules and the extensive wiring harnesses required. This has led to the introduction of centralized architecture as the next evolutionary step in E/E systems [11].

Zonal architectures, on the other hand, organize vehicle functions based on physical location rather than functional domains. Vehicle functions are implemented as software applications running on central processors, enabling shared use of sensors and actuators. This approach simplifies wiring harnesses, enhances vehicle adaptability, and decouples hardware and software, allowing for secure over-the-air (OTA) updates and greater design flexibility. The consolidation of processing units also enhances silicon integration, improving power efficiency. However, zonal architectures bring several challenges: they require high computing power and bandwidth, pose significant security concerns, and demand careful management to prevent interference between applications of varying criticality [12].

Zonal architectures also disrupt the traditional automotive value chain. The traditional value chain shows a clear vertical dependency. Original equipment manufacturers (OEMs) primarily depend on their direct suppliers, known as

1st-tier suppliers, for components and systems, while 1st-tier suppliers manage relationships with 2nd-tier suppliers, such as semiconductor manufacturers. Although OEMs have occasionally interacted directly with 2nd-tier suppliers, these engagements have been limited to conceptual discussions rather than operational dependencies.

With the emergence of "software-defined vehicles", OEMs are increasingly taking control of E/E architectures, resulting in a shift from a linear value chain to a complex web of interdependencies across multiple levels, as illustrated in Figure 1.2. This shift is forcing suppliers to rethink their roles and, in many cases, adapt their business models. Vehicle manufacturers are now demanding greater control over system and application software, aiming to reduce their reliance on 1st-tier suppliers [13].

Despite its potential, the adoption of zonal architectures is seen as a disruptive change for OEMs and will likely be introduced gradually. In the near term, a hybrid architecture appears to be a practical solution. This "mixed" architecture combines elements of traditional domain controllers and zonal structures. Domain controllers will evolve into powerful central car computers, while zonal controllers will manage body and comfort applications. Over time, functions such as powertrain and ADAS may also shift to zonal structures in later implementation stages. This transitional approach helps address critical concerns such as security, signal latency, and software legacy, balancing innovation with practical constraints [14].

In addition to academic research, the industry is actively implementing zonal architectures. For example, Tesla transitioned from a traditional E/E architecture in the Model S/X to a zone-oriented three-domain E/E architecture in the Model 3/Y, reducing the wiring harness length from 3000 m to just 1500 m or less. Similarly, BYD and other OEMs, including Volkswagen, GM, and BMW, are adopting new E/E architecture topologies to enhance efficiency and reduce complexity. Tier 1 suppliers like Bosch, Continental, Aptiv, and Huawei are also driving innovation by developing platforms and core components tailored to these advanced architectures [4]. By 2030, experts predict that vehicle E/E architectures will feature centralized computing platforms, zonal control systems, and high-bandwidth in-vehicle communication networks. The market for automotive compute units—including distributed ECUs, domain

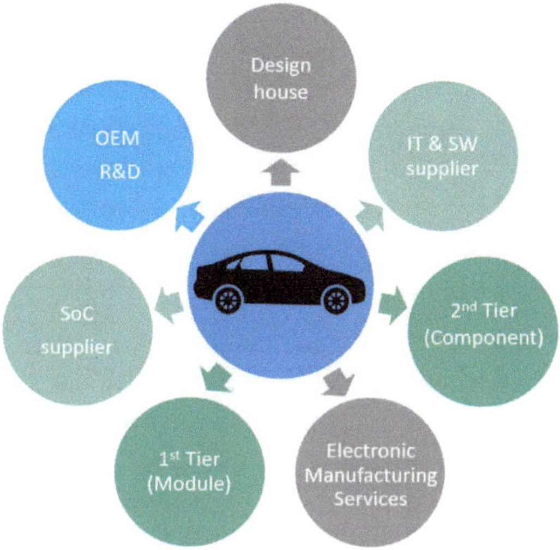

Figure 1.2: Potential automotive value chain [13]

control units (DCUs), zonal compute units (ZCUs), and central compute units (CCUs)—is projected to grow at an annual rate of approximately 6% between 2023 and 2030. While the demand for traditional ECUs is expected to decline due to functional consolidation, markets for DCUs, ZCUs, and CCUs are anticipated to grow by 30–40% annually, reflecting the industry's shift toward more integrated and software-driven systems [15].

1.2 Functional Safety

The aforementioned technological innovations are giving rise to the demand for functional safety. Functional safety refers to the systematic process of ensuring that electrical, electronic, and programmable electronic systems function safely and reliably under both normal and fault conditions, with the primary aim of preventing accidents or harm that could arise from system malfunctions. This is particularly critical in industries where failure could result in catastrophic outcomes, such as in automotive, aerospace, medical devices, and industrial machinery. By identifying potential hazards early in the design process, assessing associated risks, and implementing mitigative measures, functional safety ensures that systems remain safe throughout their lifecycle [16]. Given the increasing reliance on complex electronic and software-based systems in modern technology, the importance of functional safety has grown substantially. Failure to adhere to functional safety standards can lead to severe consequences, including loss of life, legal repercussions, and reputational damage.

Various industries follow distinct functional safety standards tailored to their specific needs. The **IEC 61508** standard serves as the umbrella for functional safety across all industries, particularly addressing electrical, electronic, and programmable electronic systems. It employs safety integrity levels (SILs) ranging from SIL 1 (lowest risk) to SIL 4 (highest risk), providing a framework for assessing and mitigating risk in safety-critical systems. This standard also forms the basis for several industry-specific standards [17]. In the automotive industry, **ISO 26262** is the key standard that governs functional safety for electrical and electronic systems in production vehicles. This standard utilizes automotive safety integrity levels (ASILs) ranging from ASIL A (lowest risk) to ASIL D (highest risk), determining the safety measures required based on the risk level of each component [18]. Notably, **SOTIF** (safety of the intended functionality) complements ISO 26262 by addressing safety concerns in autonomous driving systems and advanced driver assistance systems (ADAS), ensuring that these systems operate safely under normal conditions [19].

The aerospace industry adheres to the **DO-178C** standard, which provides guidelines for software considerations in airborne systems and equipment certification. It defines design assurance levels (DALs), which classify safety requirements based on the severity of failure conditions, ranging from DAL E (lowest criticality) to DAL A (highest criticality). This standard ensures the safety and reliability of software components in critical aviation systems, where failure can have catastrophic consequences [20].

In the railway industry, **EN 50128** establishes safety requirements for electronic systems used in railway control and protection applications. It uses software safety integrity levels (SSILs), ranging from SSIL 0 (lowest) to SSIL 4 (highest), to ensure reliable operation of rail control systems [21]. Similarly, in the medical device industry, **IEC 62304** focuses on software lifecycle processes, employing safety classifications (Classes A–C) to define requirements based on risk to patients [22]. The machinery industry follows **IEC 62061**, which applies to the safety of control systems in machinery, using SILs to manage risk [23].

This work primarily focuses on ISO 26262, the functional safety standard tailored to the automotive industry. Given the increasing complexity of modern vehicles, which rely heavily on electrical and electronic components, ISO 26262 plays a crucial role in ensuring that these systems operate safely, even in the event of malfunctions. This standard provides comprehensive guidance for managing functional safety throughout the lifecycle of a vehicle, from hazard identification to system design and validation. The importance of ISO 26262 is expected to grow as vehicles continue to incorporate more advanced technologies such as autonomous driving and smart connectivity. Chapter 2 will provide a detailed examination of ISO 26262, exploring its methodologies, requirements, and its impact on the automotive industry's approach to functional safety.

1.3 Scope of this Thesis

This thesis focuses on designing a framework for the automated exploration of E/E architecture in the automotive domain. The growing complexity of hardware and software components, combined with diverse safety-related and technical requirements, has led to a highly intricate and challenging design process. To the best of our knowledge, this process is currently handled manually by multiple development teams, each focusing on different aspects of the vehicle. Such a manual approach is not only time-consuming but can also lead to sub-optimal solutions.

To address these challenges, the proposed research investigates two central questions:

1. How can the allocation of mixed-criticality functions to hardware resources be optimized?

2. How should messages be routed and scheduled when ASIL decomposition is applied, and how can timing requirements be satisfied?

The goal is to bridge the gap between E/E architecture development and functional safety analysis by automating key aspects of the design process. The resulting framework enables the automated distribution of automotive applications across a given hardware platform. For validation and realism, the architecture of the UNICARagil project [24] is employed as a representative case study. However, the framework is not limited to this specific architecture and can be applied to a wide range of systems—ranging from distributed to centralized architectures—provided that the processing units are generic and capable of executing the assigned applications within constraints such as capacity and memory.

This framework supports applications and hardware components with varying criticality levels, which are captured by their ASIL as defined by ISO 26262 standard. Mixed-criticality tasks can be integrated onto the same hardware element only if freedom from interference (FFI) is guaranteed. Two core algorithms are introduced in this work, each addressing one of the research questions.

Criticality-Aware Resource Allocation

This algorithm addresses the first research question by automatically mapping mixed-criticality automotive functions onto a distributed heterogeneous hardware architecture. The hardware architecture consists of multiple processing units interconnected by communication buses. Applications are modeled as sets of functional blocks (referred to as tasks throughout this work), each associated with different ASILs. A development cost function is introduced to quantify the cost of deploying each task on a hardware element with a given ASIL. These cost estimates are assumed to be provided by designers as inputs to the algorithm.

Two ASIL compliance strategies are considered. The first is the so-called "Mixed-ASIL" approach, where the hardware architecture provides the necessary separation mechanisms to support mixed-criticality applications. The second is the "Max-ASIL" approach, which mandates that all software components executed on the same hardware element must comply with the highest ASIL level among the components.

The algorithm supports ASIL decomposition to enable cost-effective implementation and seeks to determine optimal mappings. Specifically, it addresses (1) the decomposition of tasks, and (2) the assignment of tasks to processing units, while minimizing development cost and ensuring compliance with safety and integrity constraints.

The problem is formulated as an integer inear programming (ILP) model. Its performance is evaluated using commercial solvers such as Gurobi and CPLEX. Additionally, a genetic algorithm (GA)-based heuristic is developed to address this problem. The results obtained using the ILP solvers and the GA-based approach are compared and discussed to evaluate their respective performance and effectiveness.

Joint Scheduling and Routing

To address the second research question, the framework is extended with a joint scheduling and routing optimization algorithm. This algorithm simultaneously addresses task and message scheduling, as well as routing, within a unified step. As inputs, the algorithm requires a precedence graph that captures the relationships among tasks, which can be generated either explicitly or randomly. Additionally, the communication network, specifying the connections between ECUs, must be provided. The algorithm supports multiple communication buses, such as CAN and Ethernet, but is not limited to these.

Each functional task is characterized by a deadline, period, and worst-case execution time (WCET) for a given ECU and ASIL. The algorithm maps tasks to ECUs in a way that avoids contention and overlap between tasks. Additionally, the algorithm considers messages exchanged between tasks and performs routing optimization to ensure that messages are schedulable and the communication buses are not overloaded. If a task is decomposed by the previously described algorithm, the corresponding messages are duplicated. For example, if a message is sent from task A to task B, and task B is decomposed into two sub-tasks, the message is also duplicated and sent to both subtasks.

Furthermore, when the algorithm is used alongside ASIL decomposition from the earlier algorithm, requirements such as FFI must be guaranteed. In this case, the algorithm ensures that communication redundancy is available for the decomposed tasks. The primary goal of this extended algorithm is to find a solution that minimizes the end-to-end latency for a given function chain while satisfying all safety, schedulability, and communication constraints.

1.4 Contributions

Focusing on optimizing automotive E/E architectures, this thesis provides the following contributions:

- **Modeling software and hardware architectures as well as requirements:** Existing approaches from the literature for modeling software and hardware architectures are reviewed, followed by the introduction of a tailored methodology for modeling E/E architectures and their requirements using System Composer in MATLAB and the AMPL tool. Formalization of models and requirements is essential to enable the automated exploration of E/E architectures. To this end, formal representations are developed, and software and hardware models are processed to serve as inputs for optimization algorithms.
- **Mapping of functions to processors:** An ILP-based approach is introduced, supported by a comprehensive set of constraints, for the automated mapping of functional blocks to hardware components. This method computes optimal hardware topologies and their associated task and signal deployments, resulting in feasible E/E architecture configurations. All constraints, boundary conditions, and optimization objectives are integrated into a unified mixed-integer programming (MIP) model and solved simultaneously.
- **Decomposition of functions into redundant lower-ASIL tasks:** ASIL compatibility is ensured during the software-to-hardware mapping process. An ILP-based method is presented for the automatic application of ASIL decomposition, as specified by the ISO 26262 standard. This technique is particularly advantageous in scenarios where suitable hardware resources are

unavailable for high-criticality applications or when the goal is to minimize development effort and cost. The framework supports both "Mixed-ASIL" and "Max-ASIL" strategies to provide increased design flexibility.

- **Message routing and scheduling:** A communication message scheduling and routing method is proposed that supports various communication networks while providing both single and redundant paths for automotive networks. The framework is ISO 26262-compliant, accommodating requirements such as FFI and independence.
- **Evaluation:** The effectiveness of the proposed framework is assessed through multiple case studies and real-world benchmarks, with a focus on performance and scalability. Results are compared with those of established techniques, such as genetic algorithms, to highlight strengths and limitations. To further assess scalability, evaluations are extended to include randomly generated architectures. Additionally, detailed procedures for generating precedence graphs with all required parameters are provided.
- **Methods to reduce computation time:** Scalability is a significant challenge due to the exponential increase in state space with a higher number of functions. To address this, existing techniques for computation time reduction are reviewed, and a binary search algorithm is proposed to limit the state space. This reduction leads to significantly lower computation times for obtaining optimal solutions.

1.5 Content Overview

The structure of this thesis is outlined as follows:

Chapter 2 introduces the E/E architecture development process and the ISO 26262 safety standard in detail. It also reviews previous research in the area of E/E architecture exploration, provides an introduction to several state-of-the-art optimization algorithms, and concludes with an analysis of the design space and components of the E/E architecture synthesis problem.

Chapter 3 presents the system models used in this thesis, covering both application and communication network models. Additionally, it explores approaches for integrating visual diagrams into the modeling framework adopted in this work. Beyond software and hardware models, this chapter also provides a detailed discussion of development cost, reliability, and ASIL decomposition models.

Chapter 4 focuses exclusively on the resource allocation problem for mixed-criticality automotive applications. In addition to the primary ILP model developed in this work, a genetic algorithm is also proposed as an alternative solution approach. Both methods are applied to a case study, and the corresponding results are presented in this chapter.

Chapter 5 begins by discussing the concepts of scheduling and routing in automotive networks. It incorporates the requirements for routing and scheduling as specified by ISO 26262. This chapter details the proposed approach to automatically route messages in the network and calculate task and message schedules. The performance of the proposed algorithm is evaluated using a case study, with results presented in this chapter.

Chapter 6 provides a quantitative analysis and evaluates the proposed methods. Several experiments are conducted to demonstrate the scalability and optimality of the work across various case studies and real-world benchmarks. Additionally, the limitations of the proposed approach are discussed, and methods to reduce computational time are explored, offering insights into enhancing the efficiency of the solution framework

Finally, Chapter 7 concludes this thesis by summarizing the key contributions and providing an outlook for future research directions.

As the content of this thesis spans multiple research domains, a dedicated literature review section is included in each chapter separately to contextualize the relevant work.

2 E/E Architecture Design and Safety Standards

The previous chapter provided an overview of current trends in the automotive industry, highlighting the necessary transformations in E/E architecture design and emphasizing the importance of adhering to safety standards. This chapter examines the development process of automotive E/E architectures, with a particular focus on ISO 26262, the primary safety standard that governs the design and development of these architectures. Furthermore, this chapter reviews state-of-the-art methodologies for E/E architecture exploration as discussed in the literature, offering a critical analysis of their advantages and limitations.

2.1 E/E Architecture Development Process

The transformations in E/E architectures and the increasing share of software components in modern vehicles have brought a surge of effort in rethinking the E/E architecture development process. Traditional development processes often follow the V-Model, which, while structured and systematic, do not fully address the new requirements posed by these advancements. The applied development strategies, tools, and methods have also shown significant gaps in adapting to this context [25]. Figure 2.1 shows the automotive development process according to the V-Model [26, 27]. The V-Model features a development phase on its downward slope, where system requirements are refined into detailed designs, and a validation phase on its upward slope, where each element is tested against corresponding specifications. The phases include system requirements, system design, detailed design, implementation, unit testing, integration testing, and system testing, with OEMs responsible for requirements, design, and final testing, while suppliers focus on implementation and unit testing [28].

While the V-Model offers advantages such as clear structure, traceability, strong validation focus, and systematic handling of multi-domain systems, it falls short in areas like adaptability, handling late-stage integration risks, and supporting

D. Mohammad Zaheri, *Safety-Oriented Optimization of Automotive E/E Architectures*, Wissenschaftliche Reihe Fahrzeugtechnik Universität Stuttgart,
https://doi.org/10.1007/978-3-658-53045-7_2

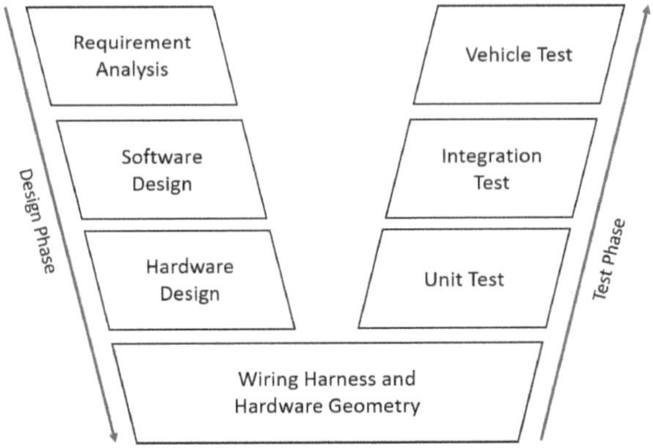

Figure 2.1: Automotive development process according to the V-model [27]

iterative or multi-speed development [29]. These limitations are particularly evident in the era of software-defined vehicles, where continuous development and integration are critical. Consequently, model-based systems engineering (MBSE) has emerged as a key methodology to address the growing challenges of managing system complexity, enabling engineers to design, validate, and optimize systems more effectively [30]. It replaces traditional document-driven processes with model-driven approaches, significantly enhancing collaboration, traceability, and consistency across development teams and domains [31]. As a result, while the V-Model continues to serve as a foundational framework, modern methodologies like MBSE have become increasingly essential to meet the demands and complexities of next-generation E/E architectures.

According to the V-Model, the design tools for vehicle development are categorized into five groups: high-level modeling, network modeling, synthesis, verification and validation, and management, each addressing specific aspects of vehicle development. High-level modeling focuses on defining system requirements and system-wide modeling, while network modeling involves the design and configuration of the network architecture and communication parameters. Synthesis emphasizes the implementation of system functionality, whereas verification and validation ensure compliance with requirements and error-free operation through diagnostics, calibration, and timing analyses. The management phase addresses acceptance validation, particularly compliance with safety standards like ISO 26262 and IEC 61508. Developing an E/E architecture is inherently complex and requires seamless data exchange across these phases. However, as noted in [28], no single tool supports all phases comprehensively, and the lack of appropriate interfaces between tools presents significant integration challenges. In this section, we provide an overview of available tools and frameworks for each category. Since E/E architecture synthesis, particularly addressing deployment problems through design space exploration (DSE) approaches and optimization objectives, is the primary focus of this thesis, a more detailed discussion of the methods and tools specific to this area will be presented in Section 2.3.

PREEvision by Vector is a comprehensive, model-based development tool that supports the entire E/E development process, from architecture design to wiring harness development. Adhering to established automotive standards, it enables the design, visualization, and optimization of E/E systems through an integrated and holistic approach [32], [33]. PREEvision incorporates various modeling layers, including requirements, logical function architecture, AUTOSAR-compliant software architecture, diagnostics, hardware architecture, wiring harness, and communication. The tool offers extensive features for requirements engineering, AUTOSAR-based software and communication design, service-oriented architecture development, and the development of safety-related systems in compliance with ISO 26262. It also includes automated consistency checks to ensure high-quality designs and supports parallel collaboration on a shared database. Furthermore, standard interfaces such as AUTOSAR, KBL, and ReqIF enable seamless data integration, promoting effective collaboration with partners and addressing the growing complexity of modern automotive E/E architectures [34].

Volcano Vehicle Systems Architect (VSA) is a powerful AUTOSAR and EAST-ADL-based design environment for E/E systems development. Built on the Eclipse framework [35], it ensures full compliance with AUTOSAR standards through its metamodel technology and provides consistency checks to support all design tasks [36]. VSA is optimized for large-scale industrial projects, offering support for multiple AUTOSAR versions and other standard formats. Key features include software component design, topology and communication design, software-to-ECU mapping, and ECU extract generation. VSA integrates seamlessly with MATLAB and Simulink, enabling round-trip data exchange for model-based application software design [37].

Rhapsody Designer by IBM is a comprehensive MBSE tool designed to support the development of complex systems and software. It provides an integrated environment for system design, simulation, and code generation, leveraging industry standards such as unified modeling language (UML) [38], systems modeling language (SysML) [39], and AUTOSAR. With features like traceability, consistency checks, and round-trip engineering, Rhapsody facilitates efficient design workflows and ensures compliance with system requirements [40].

According to [28], the tools mentioned above are commonly utilized to assist development at the high-level stage.

Several tools are available for designing and managing communication networks in automotive systems. Vector provides a toolset that facilitates the development of bus architectures, schedules, and communication matrices for CAN, LIN, and FlexRay networks [41]. Additionally, Elektrobit offers the **EB tresos**, which enables configuration, validation, and code generation for AUTOSAR basic software. Within the EB tresos product line, EB tresos AutoCore and EB corbos AdaptiveCore provide comprehensive AUTOSAR communication services, seamlessly integrating support for CAN, CAN FD, LIN, FlexRay, and Ethernet into the communication stack. EB tresos AutoCore is specifically tailored for the development of automotive ECUs in mass production projects and supports both single-core and multicore architectures, making it suitable for a wide range of communication requirements in modern automotive systems [42], [43].

ASCET-DEVELOPER by ETAS is a commercial model-based development tool designed for the efficient creation of embedded software for automotive ECUs. It supports the development of software using graphical and textual models with a focus on real-time, safety-critical applications. ASCET ensures compliance with automotive standards like AUTOSAR and ISO 26262, offering features for code generation, validation, and testing, which streamline the software development process [44].

AUTOSAR Builder is a comprehensive tool suite designed for the development and configuration of AUTOSAR-compliant automotive distributed embedded systems. It is built on the Eclipse platform with the Autosar development environment (Artop). It supports the entire workflow, including system design, software component modeling, ECU configuration, and validation. The tool ensures seamless integration with AUTOSAR standards, enhancing efficiency and collaboration in embedded system development. Additionally, its integration with EB tresos Studio, driven by a strategic partnership between Elektrobit and Dassault Systèmes, focuses on improving tool interoperability through R&D, documentation, and product support, enabling seamless collaboration and further optimization in AUTOSAR-based development processes [45], [46].

ChronSIM/ChronVAL is a tool for timing analysis of embedded systems, enabling model-based real-time simulation (ChronSIM) and formal analysis (ChronVAL). ChronSIM provides real-time simulation and optimization by using integrated hardware and software models to predict timing and performance in embedded systems [47]. ChronVAL focuses on analyzing and validating safety-critical embedded systems by calculating metrics like response times, latencies, and resource utilization [48]. Together, they support many stages of the the development process from abstract models to complete C-code implementations, ensuring robust dynamic behavior in distributed architectures [49].

SymTA/S is a commercial framework designed for system-level performance and timing analysis of distributed embedded architectures, with support for heterogeneous systems. The tool integrates optimization algorithms with system sensitivity analysis to enable rapid design space exploration. It facilitates budgeting, scheduling verification, and optimization for processors, ECUs, communication buses, and networks, while ensuring system reliability and safety. Leveraging formal scheduling analysis techniques and symbolic simulation, SymTA/S computes critical metrics such as WCET, end-to-end latencies, and resource utilization. In the context of automotive electronics, the tool provides support for industry standards such as OSEK, AUTOSAR-OS, CAN, FlexRay, and Ethernet, making it a comprehensive solution for performance optimization and timing analysis in complex embedded systems [50], [51].

DOORS an acronym for dynamic object-oriented requirements system, is a versatile, multi-platform, enterprise-level requirements management tool. It is widely acclaimed for its ability to capture, analyze, and manage system requirements while ensuring comprehensive traceability throughout the development lifecycle. The tool provides robust features for tracking requirement changes, maintaining version control, and ensuring alignment between system requirements and project objectives, making it an indispensable asset for managing complex development processes [52].

The aforementioned commercial frameworks provide extensive support for various aspects of E/E architecture synthesis, as well as verification and validation processes. Some of these tools, such as ASCET-DEVELOPER, ChronSIM/ChronVAL, and SymTA/S, include optimization features to enhance their synthesis results. However, they are limited in the number of optimization objectives they address and do not fully consider safety-relevant attributes as defined by ISO 26262, a critical standard for ensuring functional safety in automotive systems.

2.2 ISO 26262 Standard

Safety is a critical quality attribute of vehicles that warrants meticulous attention throughout all stages of their lifecycle. Vehicle safety can be examined from various perspectives, including passive safety, active safety, and functional safety. Functional safety specifically addresses hazards arising from the malfunctioning of E/E systems. Several factors influence functional safety, including organizational and management practices, development processes, system design, system types and technologies employed, and quality control methodologies [53].

ISO 26262, titled "Road Vehicles – Functional Safety," is a state-of-the-art standard in the automotive industry, adapted from IEC 61508 to provide guidelines for ensuring the functional safety of E/E systems. Initially published by the international organization for standardization (ISO) in 2011 and revised in 2018, the standard offers a comprehensive framework for identifying, mitigating, and managing risks associated with system malfunctions. Its goal is to ensure safety across the entire lifecycle of automotive systems [18].

The standard is structured into 12 distinct parts, as illustrated in Figure 2.2, each addressing specific phases and aspects of functional safety.

1. Vocabulary: Definitions of key terms and concepts to ensure a shared understanding.
2. Management of functional safety: Guidance on organizational and project-specific safety management practices.
3. Concept phase: Hazard analysis, risk assessment, and derivation of safety goals for the system.
4. Product development at the system level: Specification of requirements, design, and integration of the overall system.
5. Product development at the hardware level: Detailed guidelines for hardware design, implementation, and verification.
6. Product development at the software level: Processes and methods for developing safety-critical software.
7. Production, operation, service, and decommissioning: Safety assurance in manufacturing, installation, and operational phases.
8. Supporting processes: Definition of interactions and dependencies between customers and suppliers during development activities.
9. ASIL-oriented and safety-oriented analyses: Methods for addressing dependent failures, safety mechanisms, and ASIL decomposition.
10. Guidelines on ISO 26262: Recommendations for applying the standard in real-world projects.
11. Guidelines on application of ISO 26262 to semiconductors: Tailored guidance for specific use cases and examples.
12. Adaptation of ISO 26262 for motorcycles: Additional functional safety guidelines introduced in the 2018 revision for motorcycles.

The safety lifecycle outlined in ISO 26262 is based on a V-model and encompasses the key safety activities during the concept phase, product development,

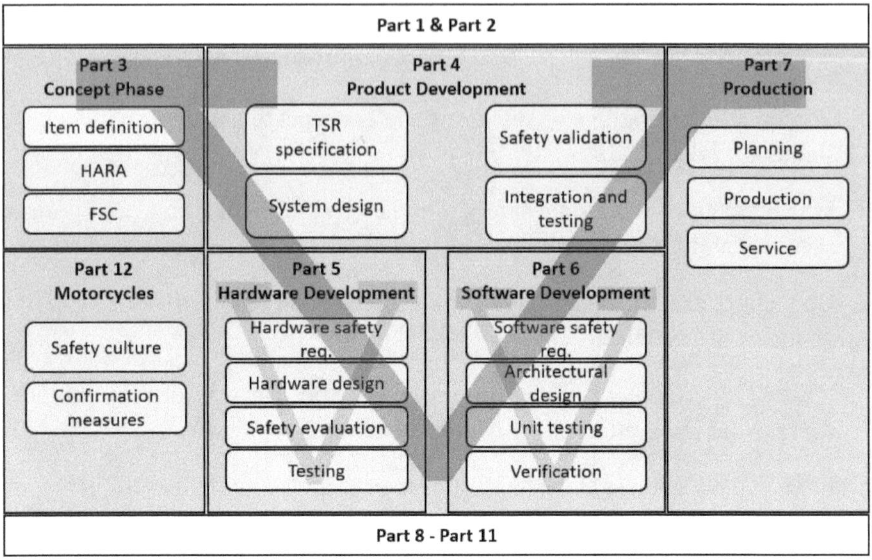

Figure 2.2: Overall structure of the ISO 26262 [18]

production, operation, and decommissioning. These phases are detailed in ISO 26262-3, ISO 26262-4, ISO 26262-5, ISO 26262-6, and ISO 26262-7. Management tasks, including planning, monitoring the progress of safety activities, and ensuring the completion of confirmation measures, are performed consistently throughout the lifecycle.

The process begins with the concept phase, where the primary objectives are to identify system hazards and define safety goals. During this phase, an "item"—a system or combination of systems—is defined for analysis. Hazard analysis and risk assessment (HARA) is conducted to identify potential hazards and classify them based on three parameters: the severity, which represents the potential harm to individuals; the exposure, indicating the likelihood of occurrence; and the controllability, reflecting the ability to avoid or mitigate the hazard. Based on these parameters, hazards are assigned an automotive safety integrity level (ASIL), which determines the rigor of safety measures required. The ASIL levels include:

- QM (Quality management): Hazards requiring no specific safety measures.
- ASIL A: Low safety requirements.
- ASIL B: Moderate safety requirements.
- ASIL C: High safety requirements.
- ASIL D: Highest safety requirements, typically for the most critical scenarios.

Table 2.1 provides a comprehensive overview of ASIL definitions based on the risk parameters: severity (S), exposure (E), and controllability (C). Each identified hazard results in the definition of a safety goal (SG), which serves as a top-level functional objective to mitigate the associated risks. Safety goals inherit the ASIL of their corresponding hazards. Functional safety requirements (FSRs) are then derived from these safety goals and outline the specific safety measures and mechanisms required to achieve them. These FSRs are detailed within the functional safety concept (FSC) and are implemented in the item's architectural elements. The ASIL assigned to a safety goal is also inherited by its corresponding FSRs, ensuring a consistent safety level throughout the system [54].

The next phase focuses on system-level development, integration, and validation. In this step, FSRs are further refined into technical safety requirements (TSRs), which are then allocated to hardware and software elements to ensure compliance at all levels of development.

Parts 5 and 6 of ISO 26262 detail smaller V-cycles for hardware and software development, respectively. These phases focus on deriving detailed safety requirements from TSRs and assigning them to specific subsystems or components for implementation. Part 7 concludes the process by covering the release of the system for production [55].

ISO 26262 emphasizes traceability throughout the safety lifecycle, linking hazards to safety goals, requirements, and implementation. This systematic approach ensures that safety objectives are met while allowing flexibility for innovations such as redundancy, architectural optimization, and automation in safety analysis.

Table 2.1: ASIL determination based on the risk parameters [18]

		C1	C2	C3
S1	E1	QM	QM	QM
	E2	QM	QM	QM
	E3	QM	QM	A
	E4	QM	A	B
S2	E1	QM	QM	QM
	E2	QM	QM	A
	E3	QM	A	B
	E4	A	B	C
S3	E1	QM	QM	A
	E2	QM	A	B
	E3	A	B	C
	E4	B	C	D

2.2.1 ASIL Decomposition

The ISO 26262 standard introduces a structured methodology known as ASIL decomposition, which facilitates the tailoring of ASIL during the concept and development phases. This approach is particularly useful when the architecture includes sufficiently independent and redundant elements, enabling the allocation of a safety requirement to two (or more) of these elements. Through ASIL decomposition, higher ASILs assigned to a safety requirement can be divided into lower ASILs for specific subsystems or components, provided that the redundancy and independence of the elements ensure equivalent safety performance. This enables designers to simplify the development and verification efforts, as lower ASIL levels demand less stringent testing, validation, and quality assurance [56]. Consequently, this approach can lead to significant cost and resource savings without compromising overall system safety. ASIL decomposition can be applied to various types of requirements, including functional safety requirements, technical safety requirements, and hardware and software requirements. ASIL defines the safety activities to be carried out

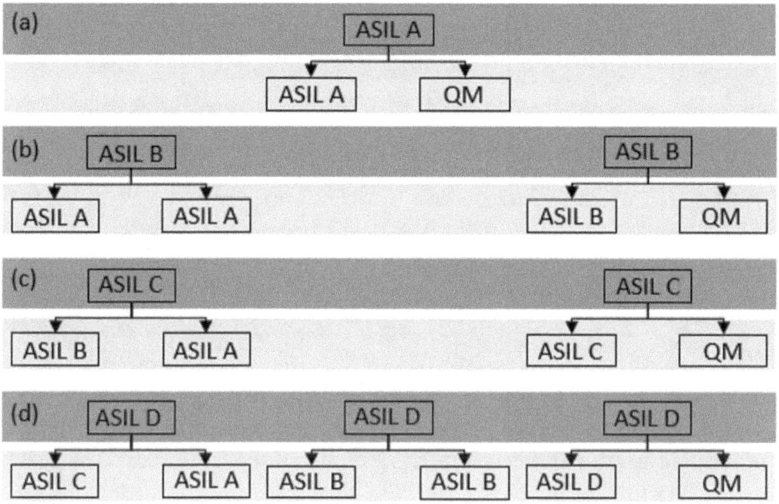

Figure 2.3: ASIL decomposition schemas in ISO 26262 [18]

throughout the safety lifecycle to maintain risk at an acceptable level. As a result, the ASIL decomposition of FSR plays a crucial role in the subsequent steps, especially during the implementation phase [55]. Figure 2.3 illustrates the ASIL decomposition schemas introduced by ISO 26262. In total, for an ASIL D requirement, five decomposition options can be used: ASIL C + ASIL A, ASIL B + ASIL B, ASIL B + 2 × ASIL A, 4 × ASIL A, and ASIL D + QM.

When performing ASIL decomposition, ISO 26262 mandates adherence to the following conditions:

- ASIL decomposition must be applied according to the permitted ASIL decomposition schemas outlined in Figure 2.3.
- The architectural elements involved in the decomposition must independently fulfill the safety requirements corresponding to the decomposed lower ASILs.
- The architectural elements involved in the decomposition must be independent.

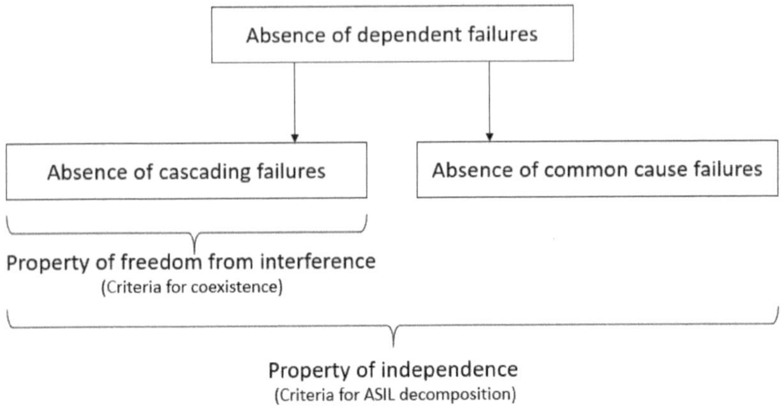

Figure 2.4: Relationship between dependent failures, FFI, and independence [18]

According to ISO 26262, independence is determined by demonstrating the absence of dependent failures. Figure 2.4 illustrates the relationship between dependent failures, freedom from interference, and technical independence. When performing ASIL decomposition, freedom from interference and the absence of common cause failures (CCFs) are used to justify independence (Clause 5, ISO 26262-9). ISO 26262 defines freedom from interference as the absence of cascading failures, which is a criterion to justify the coexistence of elements with different, or no, assigned ASILs (Clause 6, ISO 26262-9). FFI ensures that a failure in one element does not adversely affect the functionality or safety performance of another element.

According to the standard, when safety-related sub-elements implementing requirements with different ASILs, including QM, coexist within the same element, a sub-element can only be assigned a lower ASIL if evidence is provided that, for each safety requirement allocated to the element, the sub-element cannot directly or indirectly violate any safety requirement allocated to higher ASIL sub-elements. If such evidence is unavailable, the sub-element must be assigned the highest ASIL among the coexisting sub-elements for which freedom from interference cannot be demonstrated [18].

To address the implementation of a mix of safety-related and non-safety-related functionalities, ISO 26262 outlines two fundamental design strategies. The first is the "Mixed-ASIL" design, wherein functionalities with different ASILs coexist within the same system. This approach is commonly adopted when the proportion of safety-related functionalities is relatively small or when third-party or QM software needs to be integrated. The second strategy is the "Max-ASIL" design, where the entire system is developed in conformance with the highest ASIL assigned to any safety-related function within the ECU. This approach is generally preferred when the proportion of safety-related functionalities is significant [57].

The optimization algorithm proposed in this thesis is applicable to both design strategies. It facilitates the evaluation and comparison of the outcomes of these approaches, enabling the identification and selection of the optimum solution for a given system.

2.2.2 Safety Analyses

ISO 26262 necessitates a range of processes and frameworks to manage safety, develop safety concepts, and carry out verification and validation activities [58]. Qualitative methods focus on evaluating the overall design and implementation of safety mechanisms, while quantitative evaluations complement these by providing numerical evidence that safety goals are met. Quantitative analyses are used to verify hardware designs to ensure they meet targets for hardware architectural metrics and to evaluate safety goal violations caused by random hardware failures. These analyses rely on a detailed understanding of hardware failure rates and employ methods such as quantitative failure modes and effects analysis (FMEA), fault tree analysis (FTA), event tree analysis (ETA), Markov models, and reliability block diagrams [18]. The primary purpose of FMEA, and its extended version failure modes, effects, and diagnostic analysis (FMEDA), is to evaluate potential failure modes, their causes, and their effects on the system. Additionally, FMEA and FMEDA provide essential information to address safety gaps when the required hardware safety level is not fulfilled. FTA, on the other hand, is a powerful tool for identifying the possible causes of system failures. Compared to FMEDA, FTA is better suited for identifying independence requirements between architectural elements. It enables the quantification of the residual failure rate of a component failure mode and the computation of its failure probability. ISO 26262 recommends specific quantitative metrics to evaluate the safety integrity of systems, particularly in the context of hardware failures. These metrics include the single point fault metric (SPFM), the latent fault metric (LFM), and the probabilistic metric for random hardware failures (PMHF) [59]:

- **SPFM:** This metric reflects the robustness of the system against single-point and residual faults. It quantifies the proportion of faults that are either detected and mitigated or have no safety impact. A high SPFM indicates that the system effectively reduces the likelihood of single-point and residual faults in the hardware.
- **LFM:** This metric evaluates the system's robustness against latent faults by measuring its ability to tolerate undetected faults that could lead to hazardous events. A high LFM signifies a low proportion of latent faults in the hardware.
- **PMHF:** This analysis provides a quantitative assessment of the potential violation of the identified safety goal due to random failures of hardware components. It indicates the average failure probability per hour of a system on its functional lifetime. The outcome of this evaluation is then compared to a predefined target value [60].

Table 2.2 lists the target values for SPFM, LFM, and PMHF across different ASILs as mentioned in ISO 26262. It is worth noting that there is no target value for ASIL A. The table indicates that PMHF should remain below 10^{-8} failures per hour for ASIL D and 10^{-7} failures per hour for ASIL C, ensuring a high level of safety performance.

An important aspect highlighted in ISO 26262-9 is that the requirements specific to random hardware failures remain unchanged by ASIL decomposition. In other words, ASIL decomposition only impacts the measures taken to address systematic faults. Therefore, the SPFM, LFM, and PMHF remain unaffected by ASIL decomposition. Consequently, when performing ASIL decomposition, it is crucial to verify that the hardware metrics after decomposition still satisfy the requirements of the original ASIL-level for a function.

Table 2.2: Target values for SPFM, LFM, and PMHF across different ASILs [18]

	ASIL B	ASIL C	ASIL D
SPFM	$\geq 90\%$	$\geq 97\%$	$\geq 99\%$
LFM	$\geq 60\%$	$\geq 80\%$	$\geq 90\%$
PMHF	$< 10^{-7}\ \mathrm{h}^{-1}$	$< 10^{-7}\ \mathrm{h}^{-1}$	$< 10^{-8}\ \mathrm{h}^{-1}$

2.3 E/E Architecture Exploration Methodologies

Design Space Exploration (DSE) is a systematic methodology for investigating and analyzing multiple design alternatives within a predefined set of constraints and objectives. It involves evaluating various configurations and trade-offs to identify optimal solutions for specific problems. By employing computational methods and simulations, DSE facilitates the assessment of key metrics such as performance, cost, reliability, and other critical parameters associated with different design options [61].

In the context of automotive E/E architecture design, DSE plays a pivotal role in optimizing the allocation of components, wiring harness layouts, integration of diverse bus systems, and scheduling parameters. For instance, in designing zonal architectures, DSE can be employed to balance trade-offs between reducing wiring complexity, minimizing communication latency, improving power efficiency, and adhering to packaging constraints. Through the simulation and evaluation of diverse architectural scenarios, DSE enables the creation of robust, scalable, and modular E/E architectures that meet the rigorous requirements of modern vehicles, including those related to autonomous driving and electrification.

2.3.1 Deployment

As previously discussed, SDVs rely on the decoupling of software functionalities from physical hardware [62]. A crucial stage that significantly influences system quality is the mapping of software to available hardware resources, commonly referred to as resource allocation or deployment in the literature. Resource allocation can be used to ensure the optimal utilization of computational resources while adhering to critical constraints such as safety, timing, and energy efficiency, all of which are fundamental considerations in modern automotive system design. The efficient deployment directly impacts the performance, cost, reliability, and overall efficiency of the vehicle's computing infrastructure. Given the growing complexity of in-vehicle software systems, effective deployment is essential for achieving scalability, adaptability, and maintainability within SDV architectures.

Mapping techniques can be broadly categorized into design-time (static) mapping and run-time (dynamic) mapping. Static mapping is performed during the design phase and remains fixed throughout the system's operation, making it the preferred approach for applications with deterministic requirements. This method leverages comprehensive system information—such as hardware capabilities and software specifications—to determine an optimal deployment strategy. Static mapping enhances determinism and reliability by pre-assigning software functions to specific hardware components. It commonly employs graph-theoretic algorithms, mathematical programming techniques, and heuristic-based optimization to achieve a cost-efficient and constraint-aware allocation.

In contrast, dynamic mapping occurs at runtime, enabling the system to adapt to changing conditions, such as hardware failures, workload variations, or evolving computational demands. Therefore, the time required to compute a feasible and optimal solution is a crucial factor in dynamic mapping. These techniques must efficiently compute solutions under strict time constraints and are often used to minimize performance bottlenecks, balance workload distribution, and enhance system reliability.

Dynamic mapping techniques are further classified into on-the-fly mapping and hybrid mapping. On-the-fly mapping recalculates application assignments in real time whenever workload dynamics change. While this approach provides high adaptability, it may result in suboptimal solutions due to the limited processing time available for computing new allocations. To mitigate this, efficient, platform-independent heuristics are essential for improving on-the-fly mapping performance.

In contrast, hybrid mapping integrates precomputed design-time solutions with runtime adaptability. Optimized mapping solutions are generated offline and stored within the system, allowing lightweight heuristics to be applied at runtime for selecting the most suitable configuration for new applications. This approach improves mapping quality compared to on-the-fly methods; however, it has limitations, such as requiring prior knowledge of potential applications and incurring memory overhead for storing precomputed solutions.

Dynamic mapping techniques employ a range of optimization algorithms, including feedback control methods, greedy algorithms (e.g., nearest neighbor, min-max), and heuristic-based approaches (e.g., genetic algorithms, simulated annealing) [63, 64].

This thesis focuses on exploring an approach for optimizing E/E architectures using design-time mapping, where precomputed system information is utilized to ensure efficient, reliable, and constraint-aware resource allocation.

2.3.2 Related Work

This section will provide an overview of problem-solving approaches and frameworks that are applicable for synthesizing optimized design decisions. Here, emphasis is placed on both general academic and industrial approaches to E/E architecture design optimization. Additionally, a more detailed literature review, tailored to the specific content of each chapter, will be presented within the respective chapters.

In addition to the tools discussed in Section 2.1, the solutions presented in this section focus on model analysis and validation. The key distinction between these tools and those previously mentioned lies in their use of DSE techniques and their integration of optimization algorithms. These approaches enable the efficient optimization of system design, with a particular emphasis on deployment strategies.

ArcheOpterix:, introduced by Aleti et al. [65], is an open-source framework designed to optimize the architecture of embedded systems through DSE. Implemented as an Eclipse plug-in, it integrates seamlessly with the open source AADL (architecture analysis and design language) tool environment (OSATE) [66]. ArcheOpterix supports a diverse range of optimization algorithms, including exact algorithms, genetic algorithms, and ant colony optimization (ACO), while also providing extensibility for incorporating new optimization techniques [67]. These algorithms enable users to address complex trade-offs in resource allocation, timing constraints, and energy efficiency. Consequently, the framework proves highly effective for optimizing system deployment decisions, such as task allocation to processors.

Despite its advantages, ArcheOpterix exhibits several limitations. The framework is outdated and lacks comprehensive documentation, which can pose challenges to usability. Moreover, it does not support mapping analysis or address mapping problems for multi-core architectures. In addition, safety-related attributes required for compliance with ISO 26262—other than reliability—are not covered. Furthermore, the tool lacks essential features for model checking and advanced model analysis [9].

PerOpteryx:, presented by Koziolek et al., is a hybrid approach designed to optimize software architectures. It utilizes architecture models specified with the palladio component model (PCM) [68] and employs the multi-objective evolutionary algorithm NSGA-II to identify Pareto-optimal trade-offs. The framework leverages layered queueing networks (LQN) for performance analysis and can be extended to include additional tools for reliability and other quality analyses, providing flexibility for diverse evaluation needs [69]. However, PerOpteryx lacks support for defining safety parameters and does not offer capabilities for mapping analysis, limiting its applicability in certain domains.

AutoFOCUS3: is a model-based tool that supports the complete development process of embedded systems. It employs models to represent requirements, software architecture, hardware architecture, and the relationships between different viewpoints. Using these models, it provides a wide range of analysis and synthesis techniques, such as testing, model checking, deployment, and schedule generation. The tool includes a deployment viewpoint that facilitates the integration of software and hardware architectures [70]. DSE methods can be applied to determine the optimal scheduling of software components on their respective hardware components. In [71–73] the authors demonstrate how the joint generation of deployments and schedules can be efficiently performed using a DSE approach and a satisfiability modulo theories (SMT) solver. While AutoFOCUS3 proves beneficial in addressing multi-criteria deployment problems, its support for safety attributes remains limited to ASILs. Furthermore, it offers only limited optimization goals for the synthesis of E/E architectures.

SCALL: is a prototype designed to support system architects during the design phase by providing deployment solutions using DSE. It uses heuristics and the analytical hierarchy process (AHP) [74] to assist system architects in making complex allocation decisions during the early stages of design. Despite its strengths, it lacks optimization capabilities and does not account for safety-relevant constraints or requirements [75].

AAOL Framework: by Kugele et al., is a generic multi-objective evolutionary optimization tool designed for automotive E/E architecture optimization. It utilizes a domain-specific language, the automotive architecture optimization language (AAOL), to model and solve allocation problems. The AAOL framework uses DSE to find optimal deployment solutions while considering design constraints, such as memory capacity and ASIL compatibility. The objectives of the optimization are explicitly specified in order, allowing the user to define which aspects of the system, such as cost and bandwidth, should be prioritized for optimization. The multi-objective optimization mechanism targets these goals to achieve a balanced design [30]. However, the approach does not explicitly focus on redundancy mechanisms, safety attributes (only considering ASIL), or an in-depth timing analysis, which are key areas addressed in the current work.

E/E Designer: is a novel model-based framework designed for modeling and synthesizing automotive E/E architectures. It provides advanced capabilities such as automated mapping of applications, scheduling application threads across diverse hardware components, and handling message routing efficiently. The framework supports a variety of communication scenarios, including single, multicast, redundant, and homogeneous redundant routings, while also calculating schedules for communication tasks over network links. Moreover, it addresses safety requirements, such as ASIL compatibility, redundancy, and freedom from interference. The optimization of designed models is performed using multi-objective approaches that focus on key metrics, including end-to-end latency, response time, resource usage, maximum resource utilization, bandwidth consumption, link occupation rate, and cost. To achieve these goals, the framework employs mixed-integer programming (MIP) to translate requirements into constraints [76]. However, despite addressing a broad range of design-related requirements, the E/E Designer tool has notable limitations. It lacks detailed analysis and consideration of ISO 26262 safety attributes, such as the consideration of hardware metrics, and does not support ASIL decomposition. These critical gaps are key areas specifically addressed in this work.

Andromeda: is an advanced framework designed to automate the safety assessment and assurance processes for safety-critical systems [77]. It is particularly useful in the early stages of development, ensuring compliance with safety standards by streamlining assessment activities. The framework supports key safety-related tasks, including functional design and analysis, safety assessment, requirement allocation and assurance, all through automation. This makes it a valuable tool for constructing safety cases, which are crucial for certifying modern systems. At its core, Andromeda incorporates an innovative metamodel that integrates model-based safety analysis with automated and cost-efficient requirement decomposition and allocation, utilizing a metaheuristic optimization algorithm. However, despite its strengths, Andromeda supports only a limited set of objectives for optimization and does not fully account for all safety requirements that must be considered during the automation of resource allocation. Addressing this limitation is the main contribution of this work.

Numerous DSE approaches address the challenge of deploying (also referred to as allocating or mapping) software onto hardware. These methods must define constraints for valid deployments and establish optimization objectives when a simple valid deployment is insufficient. The deployment optimization problem is classified as a non-deterministic polynomial (NP) problem [78]. NP problems are those for which a proposed solution can be verified in polynomial time, but finding the optimal solution may require exponential time in the worst case. In the context of deployment, this complexity arises from the combinatorial explosion of possible mappings between tasks and processing elements, especially when considering additional constraints such as timing, safety, resource capacity, and communication overhead. As a result, many approaches rely on heuristic or metaheuristic methods, which aim to find near-optimal solutions efficiently but do not guarantee optimality [79]. Within the automotive and embedded systems domain, various approaches have been developed to tackle the deployment problem, including:

- The work in [80] introduces a model-based approach to addressing the resource allocation problem and implementing a virtualization-based environment for software deployment on the central high-performance computing (HPC) platform of software-defined vehicles (SDVs). System models are created using the Eclipse modeling framework (EMF), encapsulating detailed information about system components, as well as design decisions for resource allocation and application mapping. To facilitate automated analysis, the authors utilize EMF2SMT, a model-independent transformation technique introduced in [81], which converts user-defined models, constraints, and optimization goals into mathematical problems solvable by state-of-the-art SMT solvers such as Z3 [82]. In addition, the authors also investigate other formulation strategies, particularly an integer linear programming (ILP) formulation, to optimize resource allocation in the central computing platform of SDVs, aiming to minimize the number of virtual machines deployed on the same physical hardware under given constraints [83]. However, they do not provide any comparative results between these two methods. In contrast to the approach presented in this thesis, their approach does not consider task scheduling optimization or ASIL decomposition.

- Kampmann et al. [84] proposed an approach to automate resource allocation, focusing on optimizing power consumption and scheduling parameters. A linear model is employed to address this challenge. This study builds upon the service-oriented software architecture (SOA) presented in [85]. However, applications are only categorized as either safe or non-safe, without a detailed consideration of safety aspects. Furthermore, the approach relies solely on Ethernet as the communication protocol, without considering hybrid networking architectures that incorporate Ethernet, CAN, and FlexRay, which are more realistic in practical automotive systems.

- The authors in [86] present a comprehensive approach to address the complex problem of hard real-time allocation using constraint programming. Deployment constraints are specified to ensure that resource limitations, such as memory capacity and network usage, are not exceeded. Additionally, timing constraints for scheduling are defined. To tackle this problem, the authors propose two approaches. In the first approach, timing constraints are aggregated into a global constraint, while in the second, a decomposition method based on a logical Benders scheme is employed. This method separates the problem into a master problem, which handles allocation and resource constraints, and a subproblem that addresses timing constraints. However, the study does not focus on how the solver finds a solution.

- [87] introduces a mixed-integer linear programming (MILP)-based methodology for optimizing the deployment of software components across computing resources while considering fault tolerance and response times. The proposed MILP model accounts for the reliability and performance parameters of hardware nodes and communication links, aiming to balance these factors through a configurable trade-off. To enhance reliability, the approach replicates software components as needed and determines their optimal deployment. However, it does not take into account the criticality levels of applications, nor does it incorporate network and routing optimization.
- The authors in [88] present an integrated optimization approach for task allocation, signal packing, and the prioritization of tasks and messages in distributed automotive systems, aiming to meet end-to-end deadline constraints and minimize latencies. To handle the complexity of large-scale industrial applications, they propose a two-step synthesis method that approximates the integrated formulation. Their methodology incorporates worst-case response time analysis within a MILP framework and is evaluated against a simulated annealing-based approach for performance comparison. However, unlike the approach introduced in this thesis, their work does not account for redundancies or replications in task allocation to meet ASIL and reliability requirements. Additionally, their formulation does not include message routing optimization and, consequently, does not consider the FFI requirements specified by ISO 26262 for communication network.
- A multi-objective optimization approach is presented for the allocation of functions to control units in automotive networks. The study discusses multiple criteria as objectives of the algorithm, such as the cost of ECUs and energy consumption. Additionally, heuristic optimization algorithms are compared and improved for this specific problem. However, safety

attributes are not considered in this work and are identified as a topic for future research [89].

- [90, 91] present a design space exploration method for the automatic optimization of automotive networks. The authors argue that this optimization requires a more flexible and efficient approach than what can be achieved using either ILP or meta-heuristic approaches such as Evolutionary algorithms (EAs). To overcome these limitations, they propose a hybrid solution that combines the advantages of ILP and meta-heuristic optimization. The problem is formalized as a multi-objective 0–1 ILP, incorporating a heuristic approach based on Pseudo-Boolean (PB) solvers alongside a backtracking algorithm to address multi-objective optimization challenges. The proposed algorithm makes key decisions regarding: (1) which resources should be used in the final system (allocation) and (2) which tasks should be executed on which processor resources (binding). However, this work does not consider safety attributes, an aspect addressed in this research.
- In [92], an iterative DSE approach is proposed to support systems engineers in finding optimized solutions for various embedded systems. The authors experiment with their method on a lane-keeping system (LKS) with conflicting objectives. They consider three optimization objectives—cost, number of ECUs, and weight—and visualize the optimal solutions for each objective using a spiderweb chart to provide an intuitive solution overview for the user. However, their DSE approach does not take safety attributes or scheduling requirements into account.
- In [93], the software deployment problem is addressed with a specific focus on the requirements of the automotive domain. The study primarily concentrates on two key aspects: configuring the communication infrastructure and managing design constraints.
- In [79], a formal approach is introduced to analyze and ensure fault tolerance in automotive embedded systems by verifying whether a system design meets fail-operational requirements under various failure scenarios. The study focuses on mixed-criticality and mixed-reliability automotive systems, where functional features have different fail-operational requirements. A formal system model is developed, capturing functional features, their degradations,

software-to-hardware relationships, and deployment constraints. The approach automatically synthesizes valid software deployments, incorporating redundancy and failover mechanisms to minimize system degradation. By using formal constraints and an SMT solver, the method optimizes deployments and reconfigurations, ensuring that critical functionalities remain operational in the event of failures.

- In [5, 94], an approach for the automatic exploration of E/E architectures is presented. The proposed method is structured around three key viewpoints: E/E Architecture Viewpoint, which covers modeling aspects; Specification Viewpoint, which addresses formalization aspects; and Exploration Viewpoint, which focuses on automatic solution computation. To support this, dedicated technical meta-models and a formal language for describing E/E architecture exploration problems are introduced, along with an automatic exploration method using SMT. However, the deployment optimization in the Exploration Viewpoint—the primary focus of this thesis—does not consider safety attributes beyond ASIL compatibility as specified by ISO 26262. Additionally, the approach does not incorporate message routing or topology optimization in the exploration process.
- In [95], a model-based workflow is presented that automatically generates safety-relevant deployment constraints and optimizes deployment synthesis to fulfill safety requirements. This approach reduces development effort by streamlining the process. This work extends the existing open-source DSE implementation in the AutoFOCUS3 tool and integrates it with a Safety Pattern Synthesis tool [96]. The synthesis tool automates the recommendation of safety architecture patterns during the design of safety-critical systems. It has been implemented as a plugin in AutoFOCUS3 and provides tradeoff analysis for deployment design decisions, particularly in relation to safety and security patterns. The deployment exploration process is based on an SMT formalization and utilizes Microsoft's Z3 SMT solver as its backend. However, ASIL decomposition is not performed algorithmically to find an optimal solution; instead, it is treated as a constraint. The algorithm only verifies whether ASIL decomposition and allocation requirements are met, without actively optimizing them. Additionally, it does not cover FFI requirements.

Table 2.3 presents a comprehensive overview of the focus areas of the frameworks and approaches discussed in the reviewed literature. Additionally, it highlights the optimization algorithms utilized and their corresponding objectives. Given that safety is the primary focus of this work, the table also includes a dedicated column outlining the safety attributes considered in each approach. To the best of our knowledge, no existing work simultaneously addresses the optimization of automotive E/E architecture by providing redundancy through ASIL decomposition and optimizing network topology in a single step. This highlights a significant research gap that this work aims to fill.

Table 2.3: Analysis of optimization approaches and safety attributes in the literature

Reference	**Focus Areas**	**Optimization Algorithms/ Solvers**	**Optimization Attributes**	**Safety Attributes**
ArcheOpterix [65, 67]	Deployment and mapping	Genetic algorithm (GA), Ant colony optimization (ACO), Exact algorithms	Cost, Reliability	Reliability
PerOpteryx [69]	Allocation and feature configuration	NSGA-II	Performance, Reliability, Cost	Reliability
AutoFOCUS3 [70–73]	Model checking, Deployment, Schedule generation	SMT (Z3 solver)	Latency, Communication load	ASIL level
SCALL [75]	Software component allocation	GA	Not Applicable (N.A.)	N.A.

Reference	**Focus Areas**	**Optimization Algorithms/ Solvers**	**Optimization Attributes**	**Safety Attributes**
AAOL [30]	Deployment	Evolutionary algorithms	Cost, Weight, Bus load	ASIL level
E/E Designer [76]	Deployment	ILP (Gurobi solver)	Latency, Response time, Resource utilization, Bandwidth, Cost	ASIL level, FFI
Andromeda [77]	Functional modelling, Risk assessment, Safety analysis, Decomposition and allocation	Tabu search	Cost	ASIL level
[80, 81, 83]	Model encoding and Deployment	SMT (Z3 solver)	Number of virtual machines and utilized cores, Power consumption	ASIL level, FFI
[84]	Deployment (Service-oriented software architecture)	ILP (Gurobi solver)	Power consumption, Scheduling parameters	N.A.
[86]	Deployment and mapping	Constraint programming solver	N.A. (Constraint satisfaction problem)	N.A.

Reference	Focus Areas	Optimization Algorithms/ Solvers	Optimization Attributes	Safety Attributes
[87]	Deployment	MILP solver	Reliability, Performance	Reliability
[88]	Deployment, Signal packing, and Priority assignment	MILP (CPLEX solver)	Latency	N.A.
[89]	Deployment	HeurOpt (Heuristic optimization)	Costs, Busload, Energy consumption, and Supplier complexity	N.A.
[90, 91]	Deployment	Heuristic algorithm	Cost, Area, and Power consumption	N.A.
[92]	Deployment	Iterative multi-objective DSE	Cost, Number of ECUs, Weight	N.A.
[5, 94]	Modeling, Formalization, and Deployment	SMT (Z3 solver)	Cost, Number of ECUs, Weight, Bus load, Power consumption	ASIL level
[95]	Deployment	SMT (Z3 solver)	Cost, Latency, Reliability, Communication load	ASIL Level

2.3.3 Optimization Methods

Optimization is crucial in the design-time deployment of E/E architectures due to the presence of multiple conflicting criteria and constraints. Moreover, the vast and complex design space necessitates optimization techniques capable of efficiently exploring feasible solutions while ensuring compliance with system requirements. Without optimization, intuitive design approaches can become time-consuming and may result in suboptimal solutions.

According to [97], constraint satisfaction problems (CSPs), such as scheduling or resource allocation, which aim to find any feasible (valid) solution, are classified as NP-complete. This classification means that CSPs are at least as hard as any other problem in the NP class and cannot be solved in polynomial time. Furthermore, optimization problems, which seek an optimal solution rather than just any feasible one, are inherently NP-hard. NP-hard problems are at least as difficult as NP-complete problems but are not necessarily in NP themselves, often because verifying the optimality of a given solution may not be feasible in polynomial time. These classifications highlight the computational complexity associated with design optimization in E/E architecture synthesis, emphasizing the need for a suitable optimization algorithm to handle these challenges effectively. Most design-time methodologies adopt search-based approaches, such as genetic algorithms (GA), ILP, and simulated annealing (SA) to generate high-quality solutions. While approaches like ILP are effective in finding optimal solutions, they can face significant evaluation time challenges, particularly for large-scale problems. As system complexity increases and problems approach NP-hard characteristics, finding optimal solutions within a reasonable time frame using mathematical programming becomes infeasible. To address this, heuristic algorithms are employed to efficiently approximate solutions. While these methods do not guarantee the absolute best solution, they often provide near-optimal results with significantly reduced computational costs. Common heuristic techniques include iterative search methods like simulated annealing (SA) and tabu search, as well as population-based metaheuristics such as GA, ant colony optimization (ACO), and particle swarm optimization (PSO). In this work, GA and mathematical programming, specifically ILP and MIP, are adopted as the primary optimization methods and are briefly introduced below.

- **Mathematical Programming**

  Mathematical programming involves optimization techniques aimed at finding the best solution to a problem subject to constraints, typically formulated as equations and inequalities. The objective function, representing system goals, is optimized using methods like branch-and-bound or cutting-plane techniques. Key types include linear programming (LP), which deals with linear objective functions and constraints; integer programming (IP), where decision variables are restricted to integer values; nonlinear programming (NLP), which addresses nonlinear functions or constraints; and mixed-integer programming (MIP), which combines both continuous and integer variables. While these approaches can guarantee optimality, their computational complexity increases exponentially with problem size, limiting their practicality for large-scale systems.

- **Genetic Algorithm (GA)**

  Inspired by Darwinian evolution, the GA is a population-based optimization technique that iteratively refines candidate solutions to approach an optimal result. The pseudo-code of this algorithm is presented in Algorithm 2.1.

  GA begins with an initial population of randomly generated solution candidates, referred to as chromosomes. In each iteration, a subset of superior solutions, known as the mating pool, is selected based on a fitness function. Offspring solutions are then generated through crossover and mutation operations. The crossover operator combines two parent chromosomes to produce a new solution, while the mutation operator introduces small random variations to an individual chromosome. The rates of crossover and mutation rates determine the proportion of new solutions generated using each method. After reproduction, the combined set of current and newly generated solutions is evaluated, and the best candidates are selected for the next generation. The algorithm iterates until a termination criterion, such as a maximum number of generations or convergence to an optimal solution, is met. The final output is the best solution identified throughout the evolutionary process. The performance of GA is highly influenced by parameters such as population size, mutation and crossover rates, and selection strategies, all of which impact convergence speed and solution quality [98].

Algorithm 2.1 Genetic Algorithm (GA)

Outputs:
$\mathbf{x}^*$: Best solution found
$\mathbf{f}^*$: Best objective value

$k \leftarrow 0$
$P_k \leftarrow \{x^{(1)}, x^{(2)}, \ldots, x^{(n_p)}\}$ ▷ Initialize population
$f_k \leftarrow \text{evaluate}(P_k)$ ▷ Evaluate initial population
$(\mathbf{x}^*, \mathbf{f}^*) \leftarrow \text{best}(P_k, f_k)$ ▷ Store best initial solution
while $k < k_{\max}$ **do**
  $P_k^{\text{mate}} \leftarrow \text{select}(P_k, f_k)$ ▷ Select parents for reproduction
  $P_k^{\text{co}} \leftarrow \text{crossover}(P_k^{\text{mate}}, \alpha_c)$ ▷ Generate offspring through crossover
  $P_k^{\text{mu}} \leftarrow \text{mutate}(P_k^{\text{co}}, \alpha_m)$ ▷ Apply mutation to offspring
  $f_k^{\text{off}} \leftarrow \text{evaluate}(P_k^{\text{mu}})$ ▷ Evaluate offspring fitness
  $(P_{k+1}, f_{k+1}) \leftarrow \text{nextGeneration}(P_k, f_k, P_k^{\text{mu}}, f_k^{\text{off}})$ ▷ Select survivors
  $(\mathbf{x}^*, \mathbf{f}^*) \leftarrow \text{updateBest}(P_{k+1}, f_{k+1}, \mathbf{x}^*, \mathbf{f}^*)$ ▷ Update best solution found
  $k \leftarrow k + 1$
end while
return $\mathbf{x}^*, \mathbf{f}^*$

As resource allocation, scheduling, and routing are closely interrelated, this thesis aims to develop a unified framework that addresses them simultaneously. Given the complexity and constraint-heavy nature of the problem, it is formulated as a linear programming model. Additionally, for specific cases where these tasks can be tackled independently, a GA is proposed and its performance compared with that of the ILP approach. To further enhance computational efficiency, existing techniques from the literature are explored and their effectiveness evaluated.

3 System Model

This chapter presents the system models utilized in this thesis. A model serves as a "reduced representation" of an object [99], aiding in the understanding of a given problem. By emphasizing relevant attributes, models help avoid unnecessary complexity in the representation and analysis of the problem.

The chapter begins by introducing both application and communication network models. Additionally, It provides a detailed discussion of reliability, ASIL decomposition, and development cost models. Since the modeling approach is grounded in mathematical representation, a method is explored for systematically transforming commonly used architectural models into a corresponding mathematical framework. To facilitate this process, the MATLAB System Composer tool is investigated for creating visual diagrams that represent system architectures and enable their seamless integration into the mathematical model.

3.1 Application Model

The set of all applications in the system is denoted by A. An application can be represented as a directed acyclic graph (DAG), where its functional blocks are implemented into several computing units. Mathematically, applications are represented as a graph $G(T, D)$, where T denotes the set of tasks in G, and each $T_i \in T$ represents the i-th task of G. The edges in D indicate dependencies between tasks. The mapping of tasks to processing elements (PEs) is given by the function $X : T \rightarrow E$, where E represents the set of processing elements in the architecture.

The binary parameter $d_{i,j}$ indicates the dependency between tasks T_i and T_j. If task T_i must be executed before task T_j, then $d_{i,j} = 1$; otherwise, $d_{i,j} = 0$. The ASIL of task T_i is denoted as Λ_{T_i}.

D. Mohammad Zaheri, *Safety-Oriented Optimization of Automotive E/E Architectures*, Wissenschaftliche Reihe Fahrzeugtechnik Universität Stuttgart,
https://doi.org/10.1007/978-3-658-53045-7_3

The set of tasks requiring an ASIL level higher than any available ECU is defined as $T' \subseteq T$. ASIL decomposition constraints are applied exclusively to tasks within T'. Setting $T' = T$ subjects all tasks to decomposition. Since tasks may be decomposed, $mem_{T_i,h}$ is defined to represent the memory requirement (in megabytes, MB) of task T_i at ASIL level h.

The worst-case execution time (WCET) of T_i running on hardware component N_k with ASIL level h is denoted as $wcet_{i,k,h}$. Applications are scheduled using non-preemptive time-triggered (TT) scheduling. The period and deadline for each task T_i are given by $T_{i,p}$ and $T_{i,d}$, respectively.

3.2 Hardware Model

An automotive hardware architecture is modeled as a distributed system consisting of heterogeneous computing units interconnected by in-vehicle communication networks. The hardware model primarily comprises general-purpose ECUs, gateways, and switches. Although real-world architectures include sensors and actuators, they are not explicitly modeled, as mapping tasks to these components is not meaningful in this context. However, their influence on deployment is accounted for as a localization constraint.

The set of all processing elements (such as ECUs, processors, and cores) is defined as $E = \{E_1, \ldots, E_m\}$ and the set of network switches and gateways is denoted by $S = \{S_1, \ldots, S_l\}$. The entire set of hardware components is represented as $N = E \cup S$. Tasks are exclusively mapped to processing elements E.

The binary parameter $loc_{i,k}$ indicates the feasibility of executing task T_i on processing element E_k. If T_i can be deployed to E_k, then $loc_{i,k} = 1$; otherwise, $loc_{i,k} = 0$. The failure rate of E_k is denoted as λ_k. The ASIL level supported by E_k is represented as Λ_{E_k}. The memory capacity of E_k, expressed in megabytes (MB), is given by mem_{E_k}.

The communication overhead between tasks T_i and T_j, which applies only when they are executed on different processors, is denoted by $wcrt_{i,j}$. Note that no data transmission occurs through the communication bus when T_i and T_j run on the same processor, implying $wcrt_{i,j} = 0$.

3.3 Communication Network Model

The communication network is also modeled to support routing and message scheduling optimization. A binary parameter $l_{r,q}$ is introduced to indicate whether a physical communication link exists between two nodes N_r and N_q via a communication bus. If a link exists, then $l_{r,q} = 1$; otherwise, $l_{r,q} = 0$.

The parameter $br_{r,q}$ represents the baud rate of the communication bus connecting nodes N_r and N_q. To account for redundancy, another parameter, $rl_{r,q}$, specifies the number of redundant links between two nodes. For instance, if there are two Ethernet cables between N_1 and N_2, then $l_{1,2} = 1$ and $rl_{1,2} = 2$.

The set of all messages in the network is denoted as M. Two parameters, M_m^{src} and M_m^{dst}, indicate the source and destination tasks of message M_m, respectively. The message size, period, and deadline are represented as $M_{m,s}$, $M_{m,p}$, and $M_{m,d}$, respectively.

Since task decomposition is considered, a message originally sent between tasks before decomposition must also be transmitted between the decomposed tasks. Specifically, if message M_1 is sent from T_1 to T_2 and T_1 is decomposed into two subtasks $T_{1,1}$ and $T_{1,2}$, then M_1 must be sent from both $T_{1,1}$ and $T_{1,2}$ to T_2. If T_2 is also decomposed into $T_{2,1}$ and $T_{2,2}$, then M_1 must be sent four times:

1. From $T_{1,1}$ to $T_{2,1}$
2. From $T_{1,1}$ to $T_{2,2}$
3. From $T_{1,2}$ to $T_{2,1}$
4. From $T_{1,2}$ to $T_{2,2}$

Since task allocation and message routing are considered simultaneously, the mapping of messages depends on the mapping of their source and destination tasks to PEs. Therefore, the assignment of message M to a communication link from node r to node q is denoted by $y_{m,k,n,r,q}$, where k and n denote the nodes to which the source and destination tasks of message M are mapped, respectively, and r and q represent the nodes that form the communication link. A comprehensive overview of all notations utilized in this work is provided in the Notations section.

An exemplary mixed-criticality system with four applications is depicted in Figure 3.1a, while Figure 3.1b presents the hardware architecture comprising 11 ECUs and 4 switches, adapted from UNICARagil vehicles' four-zone E/E architecture [100]. In Figure 3.1a, both intra- and inter-application dependencies are depicted, modeling scenarios where tasks from different applications require information from each other.

3.4 ASIL Decomposition Model

In Section 2.2.1, the ASIL decomposition technique specified by the ISO 26262 standard was discussed. To formalize the decomposition schemas defined by the standard, a numerical value is assigned to each ASIL level: $A = 1, B = 2, C = 3, D = 4$. As outlined in [101], a decomposition pattern is considered valid if the sum of the ASIL values of the decomposed requirements equals the original ASIL value.

The possible decomposition schemas are derived from Eq. 3.1. The coefficient $\alpha_{i,h}$, where $h \in [1 \ldots 4]$, represents the number of redundant subtasks with ASIL h assigned to task T_i. The final values of $\alpha_{i,1}$ through $\alpha_{i,4}$ define the specific decomposition schema for each task.

$$\underbrace{\begin{pmatrix} \alpha_{1,1} & \cdots & \alpha_{1,4} \\ \vdots & \ddots & \vdots \\ \alpha_{n,1} & \cdots & \alpha_{n,4} \end{pmatrix}}_{\alpha_{i,h}} \times \begin{pmatrix} 1 \\ 2 \\ 3 \\ 4 \end{pmatrix} = \begin{pmatrix} \Lambda_{T_1} \\ \vdots \\ \Lambda_{T_n} \end{pmatrix} \quad \text{Eq. 3.1}$$

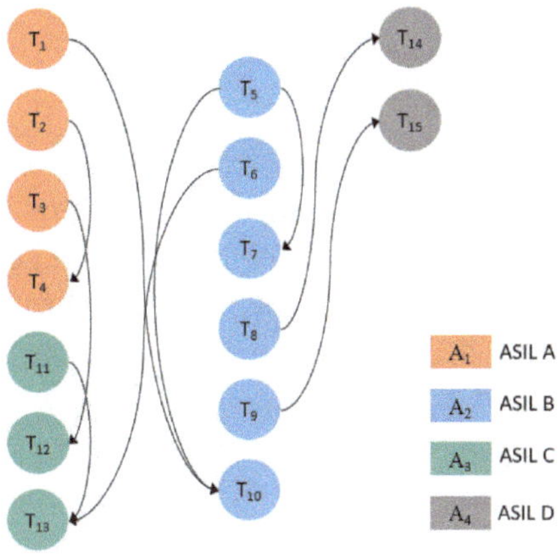

(a) Example mixed-criticality applications

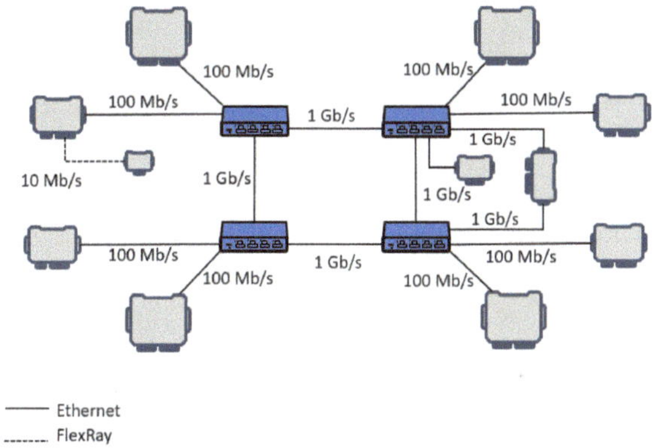

(b) UNICARagil In-vehicle network

Figure 3.1: Example software and hardware model

3.5 Reliability Model

To conduct a safety analysis, the probability of failure (PoF) is calculated and the quantitative results are compared with the target value. The probability of failure is complementary to reliability and can be calculated using Eq. 3.2 [102]:

$$\mathrm{PoF}(t) = 1 - R(t) = 1 - e^{-\lambda \cdot t} \qquad \text{Eq. 3.2}$$

In the above, λ is the constant failure rate per time unit for an ECU and t specifies the system's life time. If ASIL decomposition is applied, as defined in ISO 26262-9, the evaluation of the hardware architectural metrics and the assessment of safety goal violations due to random hardware failures remain unchanged. This means, for example, if an ASIL D task is decomposed into B(D) + B(D) and mapped to different resources, both decomposed tasks should fulfill the hardware metrics for the ASIL before decomposition, i.e. ASIL D. Eq. 3.3 is used to calculate the overall probability of failure for a task $T_i \in T'$ after decomposition:

$$\mathrm{PoF}_i = \prod_{\substack{h \in [1..4], \\ E_k \in E}} \left(1 - e^{-\lambda_k \cdot t} \cdot x_{i,k,h}\right), \quad \forall T_i \in T' \qquad \text{Eq. 3.3}$$

In this study, $t = 5000\,h$ is assumed, as specified by ISO 26262-2011 Part 10 Annex B.4. The variable $x_{i,k,h}$ is a binary decision variable and refers to the mapping of task T_i with ASIL h on E_k. Thus, a task T_i is deployed to processor E_k with ASIL h if $x_{i,k,h} = 1$. To validate the fulfillment of the PMHF requirement Eq. 3.4 is used:

$$\mathrm{PoF}_i \leq 1 - e^{-\lambda_{\mathrm{tar}}(\Lambda_{T_i}) \cdot t}, \quad \forall T_i \in T' \qquad \text{Eq. 3.4}$$

The $\lambda_{\mathrm{tar}}(\Lambda_{T_i})$ is the random hardware failure target value for task T_i with Λ_{T_i} derived from Table 2.2. It is worth noting that, due to the parallel nature of decomposed tasks, the PoF for these tasks is calculated using reliability formulas applicable to parallel systems. Eq. 3.4 represents a nonlinear constraint and hence cannot be integrated into the ILP model. The equivalent linear equation to Eq. 3.4 is derived by taking logarithms, yielding Eq. 3.5:

$$\sum_{\substack{E_k \in E \\ h \in [1...4]}} \ln(1 - e^{-\lambda_k \cdot t} \cdot x_{i,k,h}) \leq \ln(1 - e^{-\lambda_{\mathrm{tar}}(\Lambda_{T_i}) \cdot t}), \quad \forall T_i \in T' \qquad \text{Eq. 3.5}$$

Due to the fact that $x_{i,k,h}$ is a binary variable, the expression $\ln\left(1 - e^{-\lambda_k \cdot t} \cdot x_{i,k,h}\right)$ can be replaced with $\ln\left(1 - e^{-\lambda_k \cdot t}\right) \cdot x_{i,k,h}$, enabling its integration into the ILP model.

3.6 Development Cost Model

As previously stated, the decomposition and allocation of ASILs aim to ensure the fulfillment of safety requirements throughout the system development lifecycle while simultaneously minimizing the overall development cost. Consequently, the development cost model plays a crucial role in determining an optimal ASIL allocation and decomposition strategy. According to [103], various heuristic cost models have been introduced and utilized in the literature, including logarithmic and linear cost models. The cost hierarchy of ASILs is particularly important, with the assumption that ASIL D > ASIL C > ASIL B > ASIL A > QM. Furthermore, it is assumed that ASIL D > 2 × ASIL B, reflecting a more realistic cost distribution.

In this thesis, the development cost is denoted by $C_{i,k,h}$, which represents the cost of developing task T_i on processor E_k with ASIL h. While most cost models in the literature consider only task-related and ASIL-specific factors, this work recognizes that development cost is also influenced by the characteristics of the underlying hardware. To better capture this relationship, a cost model is introduced that incorporates three key attributes for more accurate estimation. Table 3.1 presents the cost models used in this work. The values provided serve as examples, and for each case study, a detailed cost table is presented in the corresponding chapter.

Table 3.1: Development cost model

Cost Heuristics	QM	A	B	C	D
Linear	0	10	20	30	40
Logarithmic	0	10	100	1000	10000
Experiential-I	0	10	20	35	50
Experiential-II	0	10	20	40	50

3.7 Representation and Transformation of System Architectures

Modeling languages play a pivotal role in graphically representing system architectures, serving as essential tools in MBSE. Standard languages such as UML and SysML offer standardized notations for specifying, visualizing, constructing, and documenting components in software and systems engineering projects. Beyond these, tools like MathWorks' System Composer [104] enable the creation of architectural models that define systems based on structure, behavior, and views. System Composer facilitates the specification and analysis of architectures, allowing for requirement mapping, architecture model optimization, and the development of designs and simulations within Simulink. Notably, it provides functionalities to export these architectures to MATLAB tables, which can then be converted to external files. This feature enables the transition of architetcural models to the proposed mathematical framework. For instance, Figure 3.2 illustrates an architecture model of a lane management system created in System Composer. By utilizing MATLAB's $find$ command, function names within this model can be extracted and used as inputs to the ILP model.

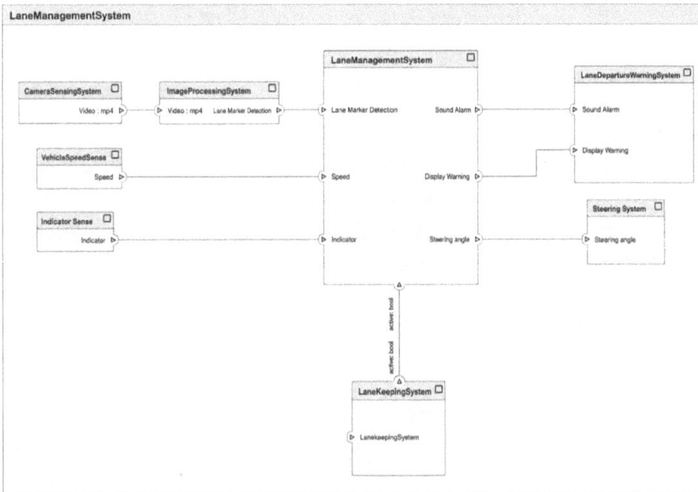

Figure 3.2: Modeling of the lane management system using System Composer

4 Criticality-Aware Design Optimization

This chapter focuses on optimizing the design of mixed-criticality automotive systems at the processor level, while deferring communication network optimization to Chapter 5. Two design strategies are examined: Mixed-ASIL and Max-ASIL.

The chapter begins by presenting the design optimization problems under consideration, illustrated with motivating examples. Subsequently, the proposed solutions are introduced, including an ILP approach and a GA method. Finally, experimental evaluations are provided, using various synthetic and real-life benchmarks to assess the effectiveness of the proposed approaches.

4.1 Problem Statement

The problem addressed in this chapter can be formulated as follows: Given a set A of applications, where each task T_i has a specified criticality level Λ_{T_i}, an architecture consisting of a set N of processing elements, the period and deadline of each task ($T_{i,p}$ and $T_{i,d}$), and the memory requirements of each task $mem_{T_i,h}$, the objective is to determine an implementation that ensures schedulability of all tasks while minimizing overall development cost and latency for critical function chains. Additionally, the solution must satisfy safety and integrity constraints.

Deriving an implementation involves the following key decisions:

1. **ASIL decomposition** of tasks, ensuring appropriate allocation of safety levels.
2. **Mapping (X) of tasks to PEs** while considering mapping constraints.
3. **Scheduling (τ) of all tasks** to ensure real-time execution within system constraints.

D. Mohammad Zaheri, *Safety-Oriented Optimization of Automotive E/E Architectures*, Wissenschaftliche Reihe Fahrzeugtechnik Universität Stuttgart,
https://doi.org/10.1007/978-3-658-53045-7_4

The problem of optimizing development cost has been actively studied in the automotive domain. Tamas-Selicean et al. [105, 106] were among the first to develop a cost model for vehicle functionalities, where they employed tabu search to optimize development cost while ensuring that all functionalities remain schedulable. Their work has inspired further research on cost optimization in mixed-criticality systems.

For instance, Gan et al. [107] explored development cost optimization using genetic algorithms. Unlike [105, 106], which incorporated a partitioned architecture, the approach in [107] does not consider partitioning. While [105, 106] contributed significantly to development cost modeling and optimization, they did not account for reliability requirements in ASIL decomposition. To address this limitation, the authors in [108] proposed two heuristic algorithms to minimize development cost while ensuring reliability through ASIL decomposition during the design phase of automotive systems. This work was further extended in [109] by introducing a two-stage solution that simultaneously addresses real-time and reliability requirements.

Our approach builds upon these studies by formulating a linear model that solves a similar problem in a single step. Additionally, the authors in [110] proposed a three-stage design process to optimize the energy consumption of automotive cyber-physical systems (ACPS) while satisfying reliability and response time requirements. Furthermore, Hu et al. [111] developed a heuristic method for generating a low-cost schedule with guaranteed safety. Their approach utilizes ASIL decomposition to construct an initial schedule with minimal development cost, followed by reliability improvements. They modeled system functionality as a directed acyclic graph and validated their approach using a real-world automotive benchmark. In [112, 113], a bottom-up and an implementation-level ASIL decomposition technique are proposed, respectively. The automotive system is described with a 3-layer model and a common-cause fault analysis is performed to validate the decomposition [114]. Dhouibi [115] introduced a methodology for ASIL allocation and decomposition through a system of linear equations, offering the capability of generating multiple solutions. While our approach to ASIL decomposition draws inspiration from this work, we diverge slightly in methodology; whereas they use fault trees to determine ASIL allocation, our work assumes pre-existing ASIL allocations to functions and focuses solely on optimizing redundancy and resource allocation using ASIL decomposition technique.

We present a simple motivational example (Fig 4.1) to illustrate and evaluate the performance of the proposed approach. As shown, the application consists of six tasks, all of which are classified as safety-critical at ASIL D. The dependencies between these tasks are represented using directed arrows, while the weights of the edges indicate the worst-case response time (WCRT).

The hardware architecture comprises four general-purpose ECUs connected via a communication bus. All ECUs are assumed to be equipped with the necessary peripherals to execute any task; hence, $\mathrm{loc}_{i,k} = 1$ for all tasks. The specifications of all ECUs are provided in Table 4.1. The memory requirements of the tasks are detailed in Table 4.2, and Table 4.3 provides the task periods and deadlines. The assumed WCETs of each tasks across different ECUs and ASIL levels are listed in Table 4.4. For a given task, the WCET is assumed to increase with higher ASIL levels compared to lower ASIL levels.

Table 4.5 presents the development costs of each task across different ECUs and ASIL levels. Specifically, the development cost of a task with ASIL D is considered greater than that of two decomposed tasks with ASIL B.

The analysis begins with a scenario in which task deadlines and periods are uniform, and the multi-objective optimization problem is solved to minimize the total execution time of the application. In the subsequent scenario, task periods from Table 4.3 are incorporated, and the corresponding results are discussed. Furthermore, Section 4.3 examines the impact of different development cost models on the candidate solutions.

Table 4.1: Properties of exemplary ECUs

Property	ECU1	ECU2	ECU3	ECU4
Λ_{E_k}	C	B	B	C
m_{E_k}	8 GB	8 GB	2 GB	16 GB
$\lambda_k[\mathrm{h}^{-1}]$	10×10^{-7}	20×10^{-7}	30×10^{-7}	8×10^{-7}

Table 4.2: Required memory for tasks at different ASIL levels

ASIL	T_1	T_2	T_3	T_4	T_5	T_6
A	2 GB	2.5 GB	1 GB	500 MB	1 GB	1.5 GB
B	2 GB	3 GB	1 GB	1 GB	1.5 GB	2.5 GB
C	2 GB	3 GB	2 GB	1.5 GB	2 GB	3 GB
D	5 GB	4 GB	5 GB	3 GB	3 GB	4 GB

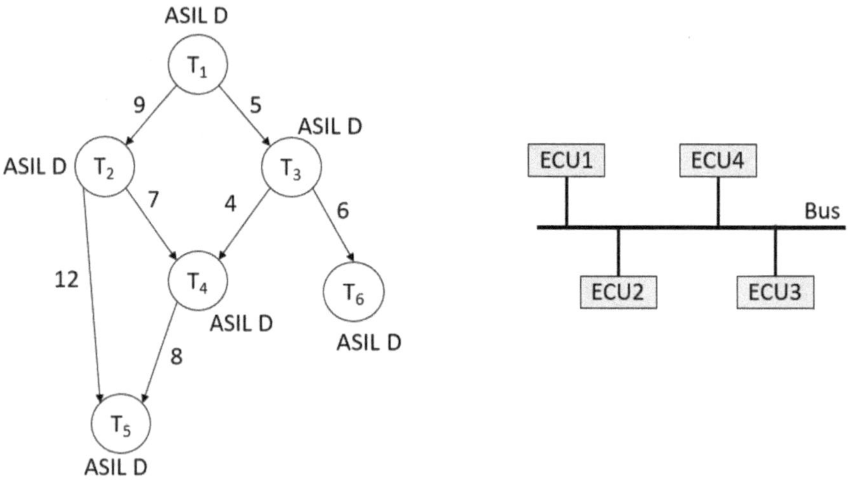

Figure 4.1: Motivational example

Table 4.3: Task specifications of the motivational example

Task	period (ms)	deadline (ms)
T_1	50	50
T_2	50	50
T_3	100	100
T_4	150	150
T_5	150	150
T_6	150	150

Table 4.4: WCETs (unit: ms) of tasks on different ECUs and ASILs

	T_1				T_2				T_3				T_4				T_5				T_6			
ASIL	E_1	E_2	E_3	E_4	E_1	E_2	E_3	E_4	E_1	E_2	E_3	E_4	E_1	E_2	E_3	E_4	E_1	E_2	E_3	E_4	E_1	E_2	E_3	E_4
A	7	4	5	8	10	6	4	7	9	8	6	13	14	11	10	10	14	12	8	16	8	1	3	8
B	9	6	7	10	12	8	6	9	11	10	8	15	16	13	12	12	16	14	10	18	10	3	5	10
C	11	8	9	12	14	10	8	11	13	12	10	17	18	15	14	14	18	16	12	20	12	5	7	12
D	13	10	11	14	16	12	10	13	15	14	12	19	20	17	16	16	20	18	14	22	14	7	9	14

Table 4.5: Development costs of tasks on different ECUs and ASILs

	T_1				T_2				T_3				T_4				T_5				T_6			
ASIL	E_1	E_2	E_3	E_4	E_1	E_2	E_3	E_4	E_1	E_2	E_3	E_4	E_1	E_2	E_3	E_4	E_1	E_2	E_3	E_4	E_1	E_2	E_3	E_4
A	8	6	5	7	9	7	7	8	5	4	3	5	4	3	2	4	6	3	3	5	8	5	5	6
B	11	8	7	9	13	9	8	12	8	6	6	8	7	6	5	7	9	5	4	8	13	9	8	10
C	18	15	12	16	20	17	17	19	14	11	11	13	12	10	8	11	15	9	8	14	22	13	13	18
D	23	17	16	21	26	22	20	25	16	14	14	17	15	15	13	15	19	13	12	18	28	22	22	22

4.2 Design Optimization Methods

Next, we propose two optimization strategies for the design problems presented in Section 4.1. In the first step, it is assumed that the system architecture provides the necessary separation mechanisms for mixed-criticality applications. Based on this assumption, a MIP model is formulated [116].

The problem is then extended by incorporating the Max-ASIL strategy, where tasks executed on the same processor must be developed according to the highest ASIL among them, resulting in increased system cost. Additionally, a GA is introduced to solve such problems and to address the limitations of the MIP model.

4.2.1 Mixed-ASIL Design Optimization

As mentioned in Section 2.2.1, according to ISO 26262, all software components executed on the same hardware element must comply with the highest ASIL among them unless evidence demonstrating fulfillment of coexistence criteria, specifically FFI, is provided. This implies that the hardware architecture must incorporate appropriate separation mechanisms to support mixed-criticality applications. ISO 26262 standard outlines various types of faults that can cause interference between software components, along with corresponding mitigation strategies. These faults include memory-related issues, which can be addressed through memory partitioning; timing-related faults, mitigated by timing monitoring; execution faults, managed through logical supervision; and interference in the exchange of information, which can be handled using end-to-end protection [117]. In the context of Mixed-ASIL design optimization, it is assumed that FFI is guaranteed by the system architecture, allowing tasks with different criticality levels to coexist on the same processor core without requiring an elevation of their ASIL classification to a higher level. Figure 4.2 illustrates an example of software tasks mapping to ECUs using this approach.

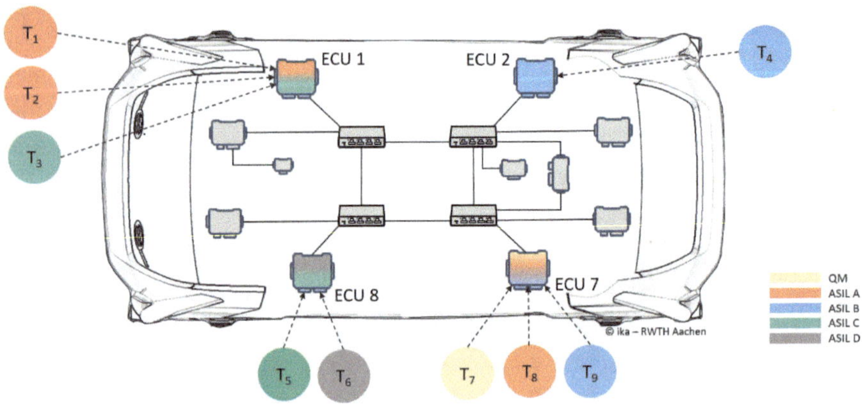

Figure 4.2: Mapping of software tasks (T_1–T_9) to ECUs according to the Mixed-ASIL design approach

A single-step multi-objective optimization problem is formulated to automate resource allocation and ASIL decomposition. The objective goals are to minimize both development cost and the maximum execution time of the application, while ensuring compliance with safety requirements. Task dependencies are taken into account, and non-preemptive schedules are computed for all tasks. The optimization objectives are formulated as follows:

$$\min \sum_{\substack{T_i \in T, \\ E_k \in E, \\ h \in [1...4]}} \mathrm{C}_{i,k,h} \cdot \mathrm{x}_{i,k,h} \qquad \text{Eq. 4.1}$$

$$\min_{A_p \in A} \Phi_p \qquad \text{Eq. 4.2}$$

Eq. 4.1 and Eq. 4.2 are used to minimize the overall development cost and the maximum execution time of a particular application ($A_p \in A$), respectively. As outlined in Chapter 3, $x_{i,k,h}$ is a decision variable, where $x_{i,k,h} = 1$ indicates task T_i is deployed on E_k with ASIL level h. The overall cost is computed as the sum of the costs associated with implementing each task T_i deployed on ECU E_k at ASIL level h.

ASIL Decomposition Schemas

Possible decomposition patterns, as defined by the ISO 26262 standard, are provided in Eq. 4.3. For example, if a task T_i with $\Lambda_{T_i} = 4$ (ASIL D) requires decomposition, one possible solution is to replace it with two redundant ASIL B(D) tasks. This is represented by $\alpha_{i,2} = 2$ and $\alpha_{i,1} = \alpha_{i,3} = \alpha_{i,4} = 0$.

$$\sum_{h=1}^{4} \alpha_{i,h} \times h = \Lambda_{T_i}, \quad \forall T_i \in T' \qquad \text{Eq. 4.3}$$

Automated Mapping

To automatically assign tasks to processing elements, the following constraint set is formulated:

$$\sum_{E_k \in E} x_{i,k,h} = 1, \quad \forall T_i \in T : T_i \notin T', h = \Lambda_{T_i} \qquad \text{Eq. 4.4}$$

$$\sum_{E_k \in E} x_{i,k,h} = \alpha_{i,h}, \quad \forall T_i \in T', h \in [1 \ldots 4] \qquad \text{Eq. 4.5}$$

$$\sum_{h=1}^{4} x_{i,k,h} \leq 1, \quad \forall E_k \in E, T_i \in T' \qquad \text{Eq. 4.6}$$

Eq. 4.4 ensures that each task, which is not subjected to decomposition, is mapped to exactly one ECU. Constraint (Eq. 4.5) is used to map decomposed tasks according to the pattern derived from Eq. 4.3. For the example provided in Section 4.2.1, this means that the two decomposed tasks with ASIL B should be mapped on two ECUs. In order to guarantee FFI and avoid deploying redundant tasks on a same ECU, Eq. 4.6 is used.

$$x_{i,k,h} = 0, \quad \forall T_i \in T, E_k \in E, h \in [1 \ldots 4] : \text{loc}_{i,k} = 0 \qquad \text{Eq. 4.7}$$

The localization constraint is applied using Eq. 4.7. It ensures that a task is mapped to an ECU only if the necessary peripheral for that task is present. Eq. 4.8 is used to guarantee that ECUs possess sufficient memory capacity to host all assigned tasks.

$$\sum_{\substack{h \in [1 \ldots 4], \\ T_i \in T}} mem_{T_{i,h}} \cdot x_{i,k,h} \leq mem_{E_k}, \quad \forall E_k \in E \qquad \text{Eq. 4.8}$$

To ensure that the CPU is not overloaded, the following constraint is used:

$$\sum_{\substack{h\in[1...4],\\ T_i\in T}} x_{i,k,h}\cdot\frac{wcet_{i,k,h}}{T_{i,p}} \leq 1, \quad \forall E_k\in E \qquad \text{Eq. 4.9}$$

where $T_{i,p}$ denotes the period of task T_i, and $wcet_{i,k,h}$ represents its execution time on processing element E_k. It is worth noting that the user can define any value less than 1 to ensure that CPU utilization remains below a preferred threshold.

ASIL compatibility is checked for generated deployments using Eq. 4.10 and Eq. 4.11, depending on whether decomposition is applied to the task or not. For clarity, the parameters Λ_{T_i} and Λ_{E_k} represent integer values ranging from 1 to 4, where ASIL D corresponds to 4 and ASIL A corresponds to 1.

$$x_{i,k,h}\cdot\Lambda_{T_i} \leq \Lambda_{E_k}, \quad \forall E_k\in E, T_i\notin T', h=\Lambda_{T_i} \qquad \text{Eq. 4.10}$$

$$x_{i,k,h}\cdot h \leq \Lambda_{E_k}, \quad \forall E_k\in E, T_i\in T', h\in[1\ldots4] \qquad \text{Eq. 4.11}$$

Constraint (Eq. 4.12) ensures the fulfillment of PMHF requirements for all decomposed tasks.

$\forall T_i\in T', t=5000:$

$$\sum_{\substack{E_k\in E,\\ h\in[1...4]}} \ln(1-e^{-\lambda_k\cdot t})\cdot x_{i,k,h} \leq \ln(1-e^{-\lambda_{\text{tar}}(\Lambda_{T_i})\cdot t}) \qquad \text{Eq. 4.12}$$

Task Scheduling

To apply the time-triggered scheduling taking the automated mapping into account, the following set of constraints is introduced [118].

$\forall(T_i,T_j)\in T; (E_k,E_m)\in E; T_i\neq T_j; E_k\neq E_m; d_{i,j}=1;\ h\in[1\ldots4]:$

$$x_{i,k,h} \cdot (\tau_{i,k} + wcet_{i,k,h}) \leq T_{i,p} \quad \text{Eq. 4.13}$$

$$\tau_{i,k} + \sum_{\substack{h_1 \in [1...4] \\ h_2 \in [1...4]}} x_{i,k,h_1} \cdot wcet_{i,k,h_1} \cdot x_{j,k,h_2} \leq \tau_{j,k} \quad \text{Eq. 4.14}$$

$$\tau_{i,k} + \sum_{\substack{h_1 \in [1...4] \\ h_2 \in [1...4]}} x_{i,k,h_1} \cdot (wcet_{i,k,h_1} + wcrt_{i,j}) \cdot x_{j,m,h_2} \leq \tau_{j,m} \quad \text{Eq. 4.15}$$

We use $\tau_{i,k}$ to denote the start time of task T_i running on E_k. Condition (Eq. 4.13) ensures that T_i completes its execution within its period slot. If task T_i must be performed before task T_j (i.e., $d_{i,j} = 1$), then the start time of T_j must be no earlier than the finish time of T_i when both tasks are allocated to the same ECU. This constraint is enforced by Eq. 4.14. If the tasks are allocated to different hardware elements, the communication time between them must also be considered. For simplicity, we assume the communication time between tasks T_i and T_j is given by the worst-case response time, denoted as $wcrt_{i,j}$. This is included in the constraints as shown in Eq. 4.15.

To ensure the overlapping-free condition for tasks T_i and T_j running on the same ECU, the following constraint must be satisfied if $d_{i,j} = 0$, i.e., there is no direct dependency between these tasks:

$$\forall (T_i, T_j) \in T; E_k \in E; T_i \neq T_j; d_{i,j} = 0; v = x_{i,k,h_1} \cdot x_{j,k,h_2};$$
$$\forall w_i \in \left[0, \frac{\text{LCM}(T_{i,p}, T_{j,p})}{T_{i,p}} - 1\right]; \quad \forall w_j \in \left[0, \frac{\text{LCM}(T_{i,p}, T_{j,p})}{T_{j,p}} - 1\right]:$$

$$\tau_{i,k} + T_{i,p} \cdot w_i + \sum_{\substack{h_1 \in [1...4] \\ h_2 \in [1...4]}} wcet_{i,k,h_1} \cdot v \leq \tau_{j,k} + T_{j,p} \cdot w_j$$

or Eq. 4.16

$$\tau_{j,k} + T_{j,p} \cdot w_j + \sum_{\substack{h_1 \in [1...4] \\ h_2 \in [1...4]}} wcet_{j,k,h_2} \cdot v \leq \tau_{i,k} + T_{i,p} \cdot w_i$$

The term LCM$(T_{i,p}, T_{j,p})$ represents the least common multiple of the periods $T_{i,p}$ and $T_{j,p}$, commonly referred to as the *hyperperiod* of the cycle. The above constraint is an either-or condition that needs to be converted to linear format. To achieve this, the Big M method is employed by introducing a binary decision variable θ_{i,j,k,w_i,w_j}. This method utilizes a sufficiently large positive constant (M) to reformulate the constraints into an equivalent linear form that can be efficiently handled by standard linear programming algorithms. The choice of M is critical to the success of the Big M method. It should be large enough to ensure that the artificial variable does not appear in any feasible solution. However, an excessively large M can lead to numerical instability and hinder the convergence to an optimal solution [119]. The above constraints are encoded in Eq. 4.17 and Eq. 4.18.

$$\tau_{j,k} + T_{j,p} \cdot w_j \geq \tau_{i,k} + \sum_{\substack{h_1 \in [1...4] \\ h_2 \in [1...4]}} wcet_{i,k,h_1} \cdot v + T_{i,p} \cdot w_i - M \cdot (1 - \theta_{i,j,k,w_i,w_j})$$

Eq. 4.17

$$\tau_{i,k} + T_{i,p} \cdot w_i \geq \tau_{j,k} + \sum_{\substack{h_1 \in [1...4] \\ h_2 \in [1...4]}} wcet_{j,k,h_2} \cdot v + T_{j,p} \cdot w_j - M \cdot \theta_{i,j,k,w_i,w_j}$$ Eq. 4.18

As discussed in [120], two periodic tasks can be executed on the same ECU if and only if their greatest common divisor (GCD) is greater than the sum of their execution times. This condition is known as the contention-free constraint and is formulated in Eq. 4.19.

$\forall E_k \in E; (T_i, T_j) \in T; T_i \neq T_j :$

$$\gcd(T_i, T_j) \geq \sum_{\substack{h_1 \in [1...4] \\ h_2 \in [1...4]}} (wcet_{i,k,h_1} + wcet_{j,k,h_2}) \cdot x_{i,k,h_1} \cdot x_{j,k,h_2} \quad \text{Eq. 4.19}$$

The maximum execution time of an application $A_p \in A$ is calculated using Eq. 4.20. This value is minimized as formulated in Eq. 4.2.

$\forall T_i \in (T \cap A_p); E_k \in E; A_p \in A :$

$$\Phi_p \geq \tau_{i,k} + \sum_{h=1}^{4} wcet_{i,k,h} \cdot x_{i,k,h} \quad \text{Eq. 4.20}$$

The constraints specified in this section can be applied to any MIP solver, such as CPLEX[1], Gurobi[2] and others, which are designed to efficiently handle linear constraints and objectives within the ILP framework.

4.2.2 Max-ASIL Design Optimization

In this section, we present our approach to designing and optimizing mixed-criticality systems without any predefined assumptions regarding FFI. In this scenario, all software components executing on the same processing element must be developed according to the highest ASIL among them. Figure 4.3 illustrates an example of software tasks mapping to ECUs based on the Max-ASIL approach. As shown, when software tasks with different ASIL levels are mapped to the same ECU, all ASIL levels are elevated to the highest ASIL assigned to that ECU. This requirement can be incorporated into the MIP model by modifying the development cost calculation.

In the Mixed-ASIL strategy, the overall cost was previously defined as the sum of the costs of each task running on a processing element, considering its corresponding ASIL. However, in this approach, we first determine the

[1] https://www.ibm.com
[2] https://www.gurobi.com

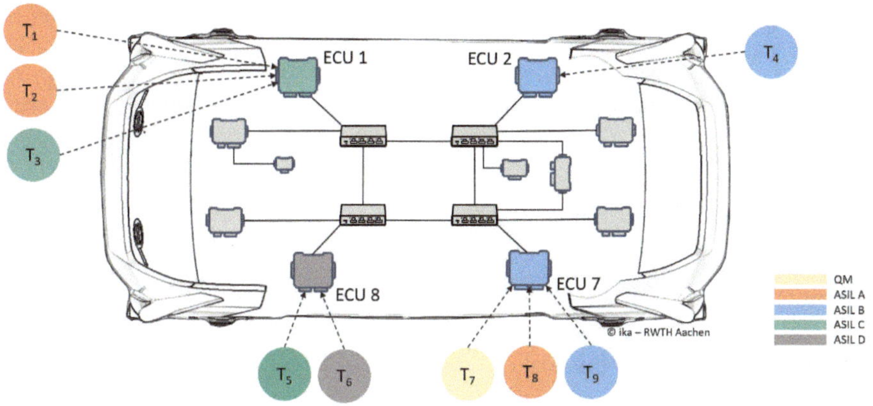

Figure 4.3: Mapping of software tasks (T_1–T_9) to ECUs according to the Max-ASIL approach

maximum ASIL level among all tasks assigned to a given hardware element. Subsequently, all tasks on that hardware element are elevated to this highest ASIL level. The overall cost is then computed as the sum of the costs of each task running on that hardware element with the determined ASIL level. To determine the maximum ASIL level among tasks assigned to an ECU E_k, Eq. 4.21 is used.

$$\beta[k] \geq h \cdot x_{i,k,h}, \quad \forall E_k \in E, \, \forall T_i \in T, \, h \in [1 \ldots 4], \qquad \text{Eq. 4.21}$$

where $\beta[k]$ is a variable that represents the maximum ASIL level assigned to the ECU E_k.

Since the approach follows linear programming principles, the value $\beta[k]$ cannot be used as an index in the cost function. Therefore, an auxiliary binary variable $\gamma_{k,o}$ is introduced to define the value of $\beta[k]$. Specifically, we express $\beta[k]$ as follows:

$$\beta[k] = \sum_{o=1}^{4} o \cdot \gamma_{k,o}, \quad \forall E_k \in E \qquad \text{Eq. 4.22}$$

with the constraint

$$\sum_{o=1}^{4} \gamma_{k,o} = 1, \quad \forall E_k \in E \qquad \text{Eq. 4.23}$$

Eq. 4.23 guarantees that only one ASIL level (i.e., one value of $\beta[k]$) is selected for each hardware element.

The overall cost is then calculated as:

$$\text{Cost}_{\text{total}} = \sum_{\substack{T_i \in T,\\ E_k \in E,\\ h \in [1...4],\\ o \in [1...4]}} C_{i,k,o} \cdot x_{i,k,h} \cdot \gamma_{k,o}, \qquad \text{Eq. 4.24}$$

where $C_{i,k,o}$ represents the cost of running task T_i on ECU E_k at ASIL level o. However, the multiplication of the decision variables $x_{i,k,h}$ and $\gamma_{k,o}$ introduces nonlinearity. To resolve this issue, an auxiliary binary variable $\delta_{i,k,h,o}$ is employed to linearize the product of the two binary variables. Eq. 4.25 - Eq. 4.27 detail the linearization procedure, ensuring that the overall formulation remains linear [121].

$\forall T_i \in T; \forall E_k \in E; h \in [1 \ldots 4]; o \in [1 \ldots 4]$:

$$\delta_{i,k,h,o} \leq x_{i,k,h} \qquad \text{Eq. 4.25}$$

$$\delta_{i,k,h,o} \leq \gamma_{k,o} \qquad \text{Eq. 4.26}$$

$$\delta_{i,k,h,o} \geq x_{i,k,h} + \gamma_{k,o} - 1 \qquad \text{Eq. 4.27}$$

Finally, the objective function, which minimizes the development cost, is formulated as follows:

$$\min \sum_{\substack{T_i \in T,\\ E_k \in E,\\ h \in [1...4],\\ o \in [1...4]}} C_{i,k,o} \cdot \delta_{i,k,h,o} \qquad \text{Eq. 4.28}$$

The other constraints introduced in the previous section will remain unchanged. To solve the multi-objective problem addressed in this thesis, the hierarchical approach proposed by Gurobi [122] is employed. This approach assigns priorities to each objective and optimizes them in descending order of priority. In this context, a higher priority is represented by a higher numerical value. For example, the following code block is used to set development cost as the higher priority objective:

```
model.setObjectiveN(Cost, 0, priority=2)
model.setObjectiveN(Latency, 1, priority=1)
```

4.2.3 Penalty-Based Genetic Algorithm

The MIP model suffers from two key limitations. First, the solving time increases as the number of tasks grows. Second, all requirements must be formulated as linear equations. However, certain requirements, such as the overall reliability of the system, cannot be expressed in a purely linear form. To address these limitations, we develop a penalty-based GA for ASIL decomposition-based resource allocation optimization. This approach incorporates reliability constraints but does not account for scheduling parameters. In [123], a GA is employed to solve the ASIL allocation problem by assigning ASIL values to the minimum cutsets (MCS) of a fault tree, using a fixed-length encoding scheme.

In contrast, our approach utilizes variable-length encoding, where each gene is assigned a random decomposition schema from a predefined library that aligns with its original ASIL level while also specifying different ECU assignments for subtasks. The code provided in Appendix A in the electronic supplementary material (ESM) demonstrates the generation of the initial population and the ECU assignment process for an ASIL D decomposition schema.

The GA begins by generating an initial population of individuals, where each individual corresponds to a possible ASIL decomposition solution. The fitness function minimizes the total cost while penalizing solutions that violate reliability constraints. Tournament selection is employed to choose parents, and crossover occurs at a randomly selected point between 1 and the length of the parent chromosome, facilitating genetic diversity. Mutation is applied by replacing a randomly selected gene with a new, ASIL-compatible decomposition schema while reassigning ECUs accordingly. To maintain high-quality solutions, elitism is applied at a rate of 0.2, preserving the top-performing individuals in each generation.

4.3 Experimental Results

This section presents the evaluation results of the proposed approaches. To this end, the MIP formulation was tested using the motivational example presented in Section 4.1. The formulated problems were implemented in Python and solved using the Gurobi and CPLEX solvers to obtain optimal solutions. The experiments were conducted on a PC equipped with an Intel i7-8565U CPU running at 1.8 GHz and 32 GB of RAM.

4.3.1 Mixed-ASIL Design Optimization

The proposed algorithm is used to solve the motivational example. In the first scenario, a simplified case is considered where no constraints are imposed on task periods. Specifically, all tasks are assumed to have the same period, which is sufficiently large to avoid any scheduling limitations. The optimal decomposition and mapping results are illustrated in Figure 4.4.

In Figure 4.4(a), the decomposition schema and mapping solution are depicted, with a focus on prioritizing development cost as the primary objective. In this instance, the overall development cost amounts to 98 $, and the maximum execution time of the application is 74 ms. Conversely, Figure 4.4(b) demonstrates the allocation scenario where latency is considered the higher-priority objective. In this case, the development cost is 109 $, and the latency is 68 ms.

As illustrated, only ECU1 and ECU4 are capable of executing tasks with ASIL C. This is consistent with the ASIL compatibility constraints and the predefined ECU specifications outlined in Table 4.1, which indicate that ECU1 and ECU4

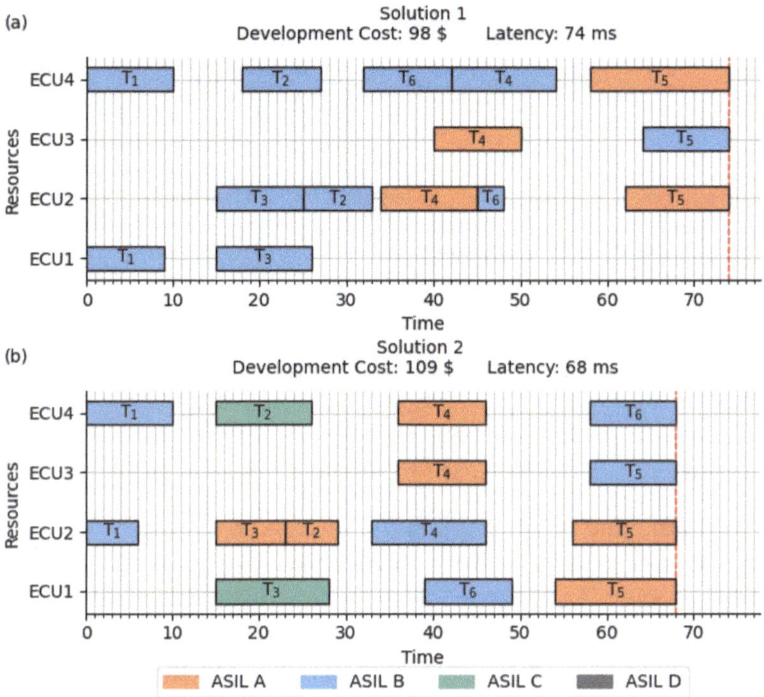

Figure 4.4: Optimal deployment solutions for the simplified case study, using Mixed-ASIL strategy. The primary objectives are: (a) minimizing development cost and (b) minimizing the maximum execution time

can support safety requirements up to ASIL C. Another crucial requirement that must be satisfied is reliability. Since all tasks must be developed in compliance with ASIL D, their failure rate should be less than 10^{-8}. Assuming an operational time of $t = 5000$ hours, the reliability goal is computed as follows:

$$R_{\text{Goal}} = e^{-\lambda \cdot t} = 0.99995 \qquad \text{Eq. 4.29}$$

Thus, the reliability of each task must exceed this threshold. After decomposition, the reliability of each task T_i is calculated using the following formula:

$$R_{T_i} = 1 - \prod_{j=1}^{n}(1 - R_j), \qquad \text{Eq. 4.30}$$

where:

- R_j is the reliability of the j-th subtask, for $j = 1, 2, \ldots, n$,
- n is the number of subtasks.

The reliability of each task in the candidate solution depicted in Figure 4.4(a) is computed as follows: $R_{T_1} = 0.999980$, $R_{T_2} = 0.999960$, $R_{T_3} = 0.999950$, $R_{T_4} = 0.999999$, $R_{T_5} = 0.999999$, $R_{T_6} = 0.999960$. Since all values exceed the reliability goal ($R_{\text{Goal}} = 0.99995$), the requirement is successfully verified.

The performance of the Gurobi and CPLEX solvers—widely acknowledged as the most commonly used ILP solvers—was compared in terms of computation time for the problem formulation. Both solvers were executed with their default configurations. The computation time of Gurobi for solving the problem with development cost as the objective was 0.266 s, whereas CPLEX took 0.844 s. When optimizing for latency, Gurobi found a solution within 7.719 s, while CPLEX required 58.188 s. Results indicate an increase in computation time across all solvers as the number of functions increases. Based on the experiments, on average, Gurobi outperforms CPLEX in terms of solving this specific problem formulation faster. As a result, the experiments conducted in this study utilize Gurobi as the solver.

In the next scenario, task periods are not uniform and are defined as specified in Table 4.3. The corresponding results are presented in Figure 4.5. As before, Figure 4.5(a) illustrates the decomposition and mapping solution when development cost is the primary optimization objective, while Figure 4.5(b) depicts the solution when latency is prioritized.

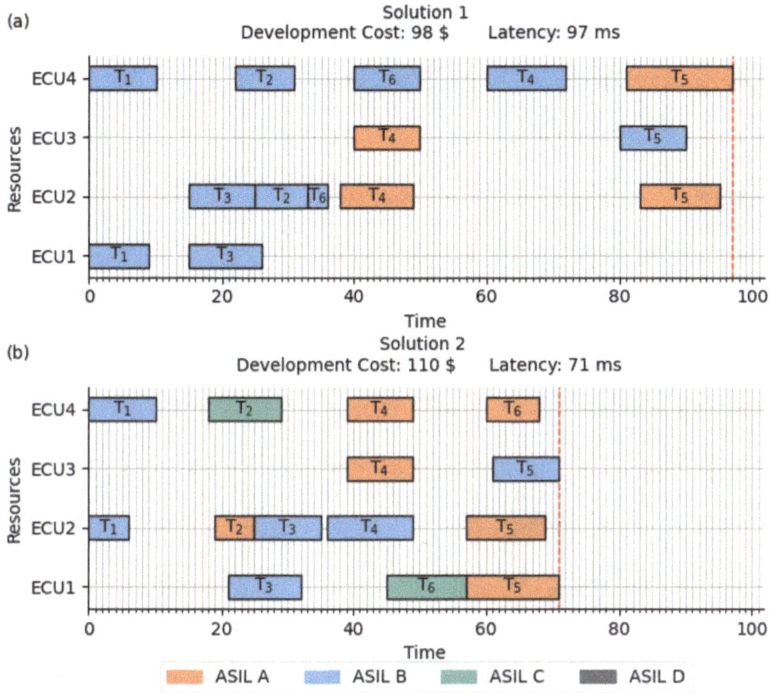

Figure 4.5: Optimal deployment solutions for the case study using Mixed-ASIL strategy with periodic tasks, considering the primary objective: (a) minimization of the development cost and (b) minimization of the maximum execution time

4.3.2 Max-ASIL Design Optimization

This section presents the results obtained by applying the proposed algorithm to optimize mixed-criticality automotive systems, without assuming any predefined conditions regarding FFI, to the motivational example. As in the previous section, for the first scenario, it is assumed that all tasks have the same large period, indicating the absence of task period constraints. The optimal decomposition and mapping results are illustrated in Figure 4.6.

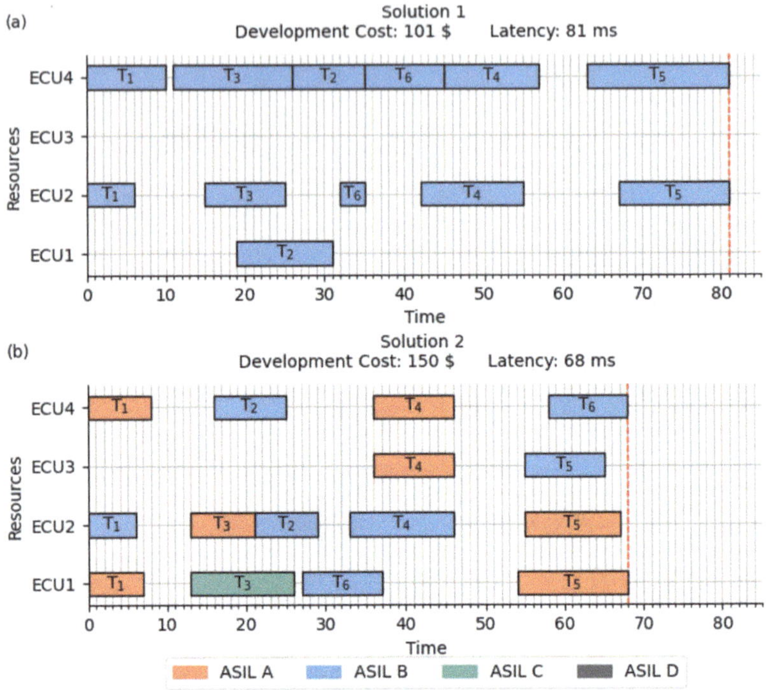

Figure 4.6: Optimal deployment solutions for the simplified case study, using Max-ASIL strategy. The primary objectives are: (a) minimizing development cost and (b) minimizing the maximum execution time

As shown, the development cost in both cases—where either development cost or latency is prioritized—is higher compared to using the Mixed-ASIL design strategy. This outcome is expected, as cost increases with higher ASIL levels. Since all tasks executed on the same processor element must be developed according to the highest ASIL level among them in this approach, the total cost is anticipated to increase relative to the previous case, where mixed-criticality tasks can coexist.

The reliability values for each task in the solution depicted in Figure 4.6(a) are calculated as follows: $R_{T_1} = 0.999960$, $R_{T_2} = 0.999980$, $R_{T_3} = 0.999960$,

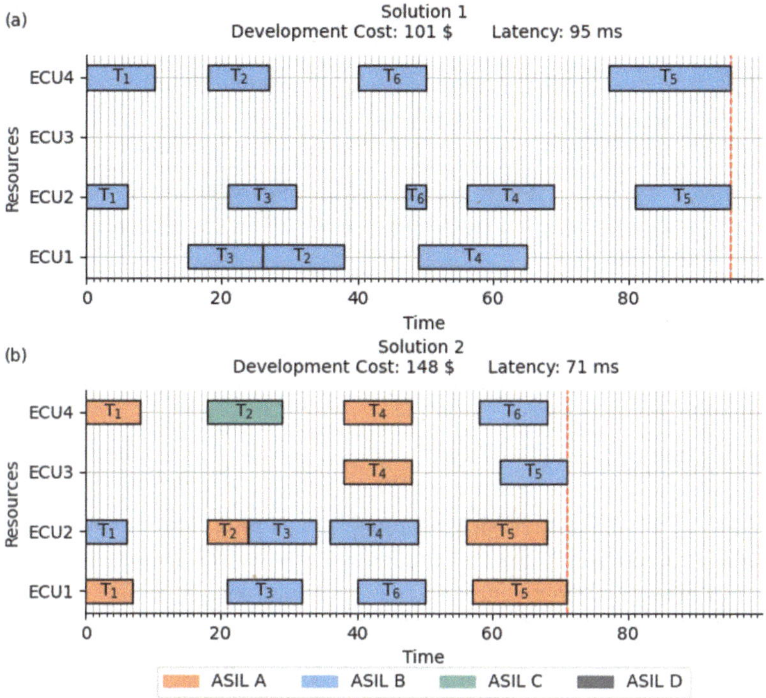

Figure 4.7: Optimal deployment solutions for the case study using Max-ASIL strategy with periodic tasks, considering the primary objective: (a) minimization of the development cost and (b) minimization of the maximum execution time

$R_{T_4} = 0.999960$, $R_{T_5} = 0.999960$, $R_{T_6} = 0.999960$. These values verify the fulfillment of the reliability requirement.

In the next scenario, the problem is solved while considering task periods specified in Table 4.3. The results are illustrated in Figure 4.7.

The required computation time for Gurobi to solve the presented multi-objective problem using the Max-ASIL design strategy is higher compared to the Mixed-ASIL strategy. Table 4.6 provides a detailed comparison of the computation time required for Gurobi to find a feasible solution under different strategies and priority settings.

Table 4.6: Computation time (in seconds) for solving the multi-objective problem using different design strategies and priority settings

Strategy	Scenario	Cost as Higher Priority	Latency as Higher Priority
Mixed-ASIL	Simplified case	0.645	12.94
	Periodic tasks	1.562	49.69
Max-ASIL	Simplified case	1.713	30.44
	Periodic tasks	6.243	72.08

4.3.3 Evaluation of Cost Heuristics

Ideally, finding an optimal ASIL decomposition and task mapping requires precise cost information for each component and its ASIL implementation. However, for newly developed elements, such cost information may be unavailable. In the literature, different cost heuristics have been defined to address such problems. For instance, [124] employs a linear cost heuristic in which each ASIL level is assigned a cost equal to its corresponding integer value as defined by the ASIL decomposition rules. This represents a relatively simplistic cost model. Instead, multiple heuristics can be formulated to evaluate ASIL costs [125]. In Table 3.1, various cost heuristics are presented. The use of different cost heuristics often results in varying optimal solutions for the same problem. This section evaluates different cost heuristics in terms of solving time and the resulting decomposition schemas.

The first cost heuristic is **linear**. In this model, the cost of developing a task T_i on ECU E_k with ASIL B is assumed to be twice the cost of developing the same task on the same hardware with ASIL A. The cost associated with ASIL C is assumed to be three times, and for ASIL D, four times that of ASIL A. Formally, the costs are defined as follows:

$$C_{i,k,2} = 2 \times C_{i,k,1}, \quad C_{i,k,3} = 3 \times C_{i,k,1}, \quad C_{i,k,4} = 4 \times C_{i,k,1}.$$

In the **logarithmic** cost heuristic, the development cost increases by a factor of 10 for each increment in ASIL level. Specifically, the costs are defined as:

$$C_{i,k,2} = 10 \times C_{i,k,1}, \quad C_{i,k,3} = 100 \times C_{i,k,1}, \quad C_{i,k,4} = 1000 \times C_{i,k,1}.$$

In **experiential-I**, it is assumed that costs increase with increasing ASIL, with the only constraint being that the cost of ASIL D must be greater than twice the cost of ASIL B. The values shown in Table 3.1 are used, such that:

$$C_{i,k,2} = 2 \times C_{i,k,1}, \quad C_{i,k,3} = 3.5 \times C_{i,k,1}, \quad C_{i,k,4} = 5 \times C_{i,k,1}.$$

For **experiential-II**, in addition to the requirements of experiential-I, it is assumed that the cost increase from ASIL B to ASIL C is greater than any other transition. In this case:

$$C_{i,k,2} = 2 \times C_{i,k,1}, \quad C_{i,k,3} = 4 \times C_{i,k,1}, \quad C_{i,k,4} = 5 \times C_{i,k,1}.$$

To ensure a fair comparison, random values between 1 and 10 are first generated for the cost of ASIL A, and the cost table for each heuristic is extended accordingly. The motivational problem is then solved by minimizing the overall cost as the primary objective, while considering various cost heuristics using both the Mixed-ASIL and Max-ASIL approaches in a simplified case. Additionally, all tasks are assumed to have the same period of 100 ms.

Figures 4.8 and 4.9 illustrate the distribution of tasks across ASIL levels under different cost models using the Mixed-ASIL and Max-ASIL approaches, respectively. As observed, when the cost model follows a logarithmic trend, tasks are more likely to be assigned to ASIL A and B. This is expected, as the cost increases by a factor of 10 with each ASIL increment, making development at ASIL C significantly more expensive and substantially raising the overall cost.

The linear and experiential-I models exhibit similar trends since the costs of development are almost the same, with the cost of developing at ASIL C and D in experiential-I being slightly higher than in the linear model. Therefore, the number of tasks assigned to ASIL C is either the same as or lower than in the linear model.

In the experiential-II model, the cost increase from ASIL B to C is significantly higher, resulting in no tasks being assigned to ASIL C. Instead, tasks tend to be assigned to ASIL A and ASIL B. Compared to the logarithmic model, the experiential-II model results in a higher proportion of tasks at ASIL B, as the cost increase from A to B is only a factor of 2, whereas in the logarithmic model, it increases by a factor of 10.

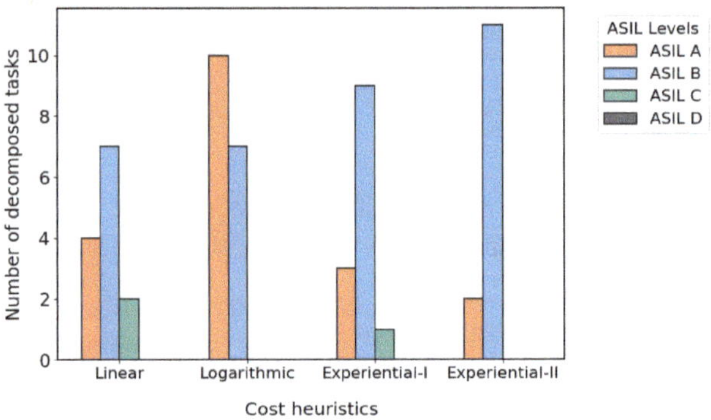

Figure 4.8: Distribution of tasks across ASIL levels under different cost models using the Mixed-ASIL approach

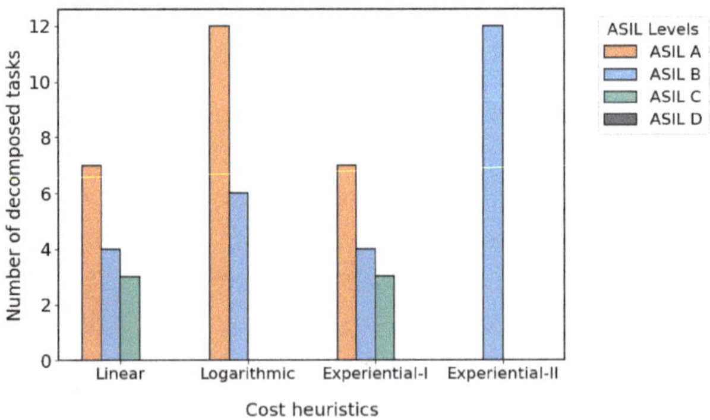

Figure 4.9: Distribution of tasks across ASIL levels under different cost models using the Max-ASIL approach

The results of applying the proposed decomposition optimization methods to the motivational example using the Mixed-ASIL and Max-ASIL approaches are presented in Table 4.7 and Table 4.8, respectively. For each cost heuristic, the optimal solution that achieves the minimum cost (higher-priority objective) and latency is identified. The column "Minimum Cost" represents the optimal cost value of the solution, while "Solving Time" lists the average processing time (in seconds).

Table 4.7: Comparison of optimization results for different cost heuristics using Mixed-ASIL approach

Cost Heuristic	Minimum Cost	Solving Time (s)
Linear	83	0.723
Logarithmic	219	0.580
Experiential-I	85	0.692
Experiential-II	86	0.661

Table 4.8: Comparison of optimization results for different cost heuristics using Max-ASIL approach

Cost Heuristic	Minimum Cost	Solving Time (s)
Linear	90	2.505
Logarithmic	372	2.155
Experiential-I	95.5	2.713
Experiential-II	96	2.711

4.3.4 Evaluation of Penalty-Based Genetic Algorithm

As previously mentioned, one of the advantages of the proposed GA is its ability to incorporate overall system reliability as a requirement. The overall reliability of a system is computed as follows:

$$R_{\text{tot}} = \prod_{T_i \in T} R_{T_i} \qquad \text{Eq. 4.31}$$

Although some studies [126, 127] have proposed methods to approximate this calculation in a linearized format, these approaches are often computationally complex and time-consuming.

To demonstrate the effectiveness of our approach, the motivational example is applied to the GA. In addition to constraints on the reliability of each decomposed task, an additional requirement is introduced to enforce a restriction on the overall system reliability. Figure 4.10 illustrates the obtained results. As the overall reliability threshold increases, the development cost also rises. Further evaluation regarding scalability and a comparative analysis with the MIP approach are discussed in Section 6.

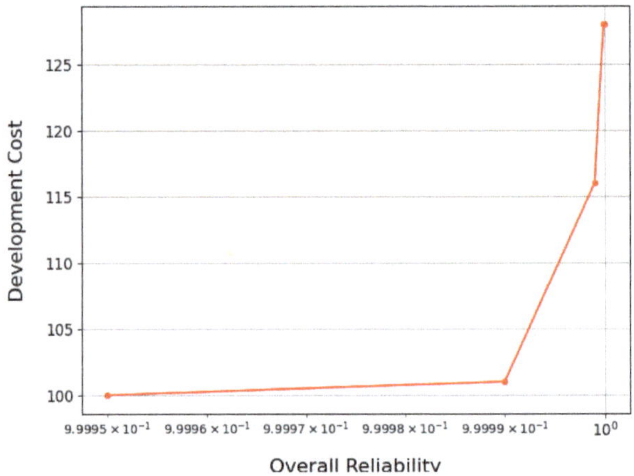

Figure 4.10: Impact of increasing the overall reliability threshold on development cost, achieved by the penalty-based GA

5 Topology and Routing Optimization

This section extends the approach introduced in Chapter 4 by incorporating design optimization at the network level. The proposed model enables simultaneous routing and resource allocation optimization in a single step. This integration is crucial, as all communication-related attributes—such as message sizes, bus load, and network topology—along with resource specifications, including task dependencies and safety levels, collectively influence the selection of the optimal solution.

The chapter begins with a literature review on existing research in E/E architecture topology and routing optimization. It then presents a MIP formulation to address the proposed problem.

5.1 Problem Statement

Modern automotive networks consist of up to hundreds of ECUs connected via various types of shared buses, such as CAN and FlexRay [128]. However, these architectures are being increasingly replaced by zonal architectures, where each zone within the vehicle manages a range of functions specific to that area. This transition streamlines the wiring system while reducing complexity, weight, and cost [129]. In zonal architectures, zonal controllers are interconnected using Ethernet, enabling a more efficient and scalable communication infrastructure. Despite these benefits, zonal architectures introduce new challenges—particularly in satisfying stringent requirements related to *latency*, *safety*, and *bandwidth*. To address these challenges, it is essential to ensure that all time-critical message deadlines are met, fault tolerance and FFI are maintained, and the communication busload remains within the permissible limits during the design of E/E architectures. These objectives can be achieved by employing a specialized optimization algorithm that is explicitly designed

D. Mohammad Zaheri, *Safety-Oriented Optimization of Automotive E/E Architectures*, Wissenschaftliche Reihe Fahrzeugtechnik Universität Stuttgart,
https://doi.org/10.1007/978-3-658-53045-7_5

to accommodate the complex constraints and interdependencies inherent in this problem.

The problem of vehicle network optimization has been addressed in various studies [130–134]. In [135] a combination of ILP and a multi-objective evolutionary algorithm (MOEA) is proposed to automatically optimize both the topology and routing in automotive networks. The optimization objectives include minimizing cost and the WCRT of all messages. To enhance efficiency, a preprocessing algorithm is introduced to reduce the search space and accelerate computation time. In their approach, it is assumed that the functions are already mapped to the ECUs during the integration phase. However, the integration of functions into ECUs and the selection of network topology can significantly influence one another. Therefore, these aspects should be considered jointly during the design process. Askaripoor et al. [136] took this research a step further by proposing an ILP model to optimize network architecture while supporting homogeneous redundant routing. This model is part of a broader framework for the design and synthesis of vehicle E/E architectures [76]. While our work also supports redundancy, we explicitly consider safety attributes as specified in ISO 26262. Furthermore, since our approach incorporates ASIL decomposition, it imposes stricter constraints compared to existing methods.

In [137], an ILP-based approach is presented to optimize the joint routing and scheduling (JRaS) problem in time-sensitive networking (TSN). The authors propose various optimizations that effectively reduce the problem's complexity, achieving an 80% reduction in runtime compared to the base ILP model.

In [138], an approach for mapping and topology optimization is presented. Unlike our approach, the mapping of functions to hardware is performed in the first step, and the result of this optimization is then used to optimize the network topology. The optimization focuses on identifying k optimal clusters such that each ECU belongs to only one cluster, while ECUs acting as gateways may be connected to more than one cluster. The objective is to minimize both the cable length and the amount of cross-cluster communication. However, this approach does not support redundancy or safety requirements.

To the best of our knowledge, only in [139] are routing and scheduling optimization addressed in conjunction with ASIL decomposition. The authors present an

SMT-based formulation that integrates ASIL decomposition, switch selection, routing, and scheduling. However, in their approach, the source and destination ECUs of each message are predefined, implying that the function-to-ECU allocation is assumed to be fixed. In contrast, our approach constitutes a more comprehensive framework that simultaneously considers mapping and routing optimization in a single step, while also incorporating ASIL decomposition and physical path redundancy.

The problem addressed in this chapter can be formulated as follows: Given a set of processing elements E, a set of network switches and gateways S, and a set of messages M, where each message M_i is characterized by its size $M_{i,s}$, deadline $M_{i,d}$, period $M_{i,p}$, source M_i^{src}, and destination tasks M_i^{dst}, the objective is to determine an optimized implementation. This implementation must ensure that all message deadlines are met, routing constraints are not violated, and communication link capacities are not exceeded.

Determining an optimal implementation involves the following key decisions:

1. **Assignment (Y) of communication messages to network links**, ensuring that routing constraints are satisfied and bandwidth requirements are met.
2. **Scheduling (σ) of communication messages**, ensuring that all timing requirements and message deadlines are fulfilled.

For this problem, multiple optimization objectives can be considered. In this work, the objective is to optimize the network configuration to ensure that all messages are schedulable while minimizing end-to-end latency.

In scenarios where the sender and/or receiver of a message are decomposed into redundant subtasks, the proposed model accordingly treats the messages as redundant and generates individual schedules for each message instance on the corresponding nodes. Furthermore, the model ensures that redundant message paths are disjoint and resilient to potential faults in the network, thereby fulfilling the independence requirements mandated by ISO 26262 for ASIL decomposition.

5.2 Design Optimization Strategy

The MIP formulations presented in this section are employed to automate the routing of communication messages across the network. These formulations account for both path and message dependencies and are capable of computing schedules for all routed communication messages.

As discussed earlier, the constraints introduced in this section can be integrated with those presented in Chapter 4 to enable joint task-to-ECU mapping and message routing in a single optimization step. Consequently, the variable $x_{i,k,h}$, which represents the assignment of task T_i to ECU E_k with ASIL level h, is also incorporated into the constraints defined in this section.

5.2.1 Automatic Message Routing

In automotive networks, message routing plays a crucial role in managing the transfer of communication data between various nodes, including ECUs, switches, and gateways. These messages carry information related to essential vehicle functions, such as engine control, braking, steering, and infotainment systems. The routing process ensures that each message is delivered to its intended destination in a timely, reliable, and efficient manner, thereby enabling the coordinated operation of distributed systems. The primary objective of message routing is to optimize data transmission while adhering to network constraints. Routing decisions are influenced by several factors, including the message's timing requirements, the network's topology, the available bandwidth of communication links, and fault-tolerance strategies, such as redundancy. As modern vehicles incorporate an increasing number of complex, safety-critical systems, efficient message routing becomes essential to ensure seamless, real-time communication across the vehicle's electronic and communication infrastructure.

The set of all vehicle nodes is defined as N, which comprises the set of ECUs, denoted by E, and the set of switches and gateways, denoted by S. Functional tasks are constrained to be mapped exclusively onto processing elements. This restriction is enforced using Eq. 5.1.

$$x_{i,s,h} = 0, \quad \forall T_i \in T, \forall N_s \in S, \forall h \in [1 \ldots 4] \qquad \text{Eq. 5.1}$$

The following set of constraints is integrated into the model to formalize and encode the message routing process:

$\forall T_i, T_j \in T; T_i \neq T_j; \forall N_k, N_n, N_r, N_q \in N; \forall M_m \in M; M_m^{src} = T_i; M_m^{dst} = T_j$:

$$\sum_{h=1}^{4} x_{i,k,h} + \sum_{\substack{N_n \in N \\ N_r \in N \\ k \neq n}} y_{m,k,n,r,k} \leq 1 \qquad \text{Eq. 5.2}$$

$$\sum_{h=1}^{4} x_{j,k,h} + \sum_{\substack{N_n \in N \\ N_r \in N \\ k \neq n}} y_{m,n,k,k,r} \leq 1 \qquad \text{Eq. 5.3}$$

$$y_{m,k,n,r,q} \leq l_{r,q}, \quad r \neq q \qquad \text{Eq. 5.4}$$

$$y_{m,k,k,k,k} = \sum_{\substack{h_1=1 \\ h_2=1}}^{4} x_{i,k,h_1} \cdot x_{j,k,h_2} \qquad \text{Eq. 5.5}$$

$$y_{m,k,k,k,k} + \sum_{\substack{N_n, N_r \in N \\ n \neq k}} y_{m,k,k,n,r} \leq 1 \qquad \text{Eq. 5.6}$$

$$y_{m,k,n,r,q} + y_{m,k,n,q,r} \leq 1, \quad k \neq n \qquad \text{Eq. 5.7}$$

$$y_{m,k,n,r,q} \leq \sum_{h=1}^{4} x_{i,k,h} \qquad \text{Eq. 5.8}$$

$$y_{m,n,k,r,q} \leq \sum_{h=1}^{4} x_{j,k,h} \qquad \text{Eq. 5.9}$$

$$\sum_{N_q \in N} y_{m,k,n,r,q} \leq 1 \qquad \text{Eq. 5.10}$$

$$\sum_{N_r \in N} y_{m,k,n,r,q} \leq 1 \qquad \text{Eq. 5.11}$$

The source and destination tasks for each message $M_m \in M$ are assumed to be given as T_i and T_j, respectively. The binary decision variable $y_{m,k,n,r,q}$ represents the assignment of messages to network links. It takes the value 1 if message M_m, whose source task T_i is mapped to E_k and destination task T_j is mapped to ECU E_n, is transmitted over the link from node N_r to node N_q.

In Eq. 5.2, the sender node E_k is required to block all incoming messages related to M_m that are coming from other nodes. Similarly, based on Eq. 5.3, the destination node is constrained to block all outgoing transmissions of the message. In other words, if $x_{j,k,h} = 1$, then $y_{m,n,k,k,r} = 0$. Here, $x_{j,k,h} = 1$ indicates that the destination task T_j is implemented on ECU E_k, and $y_{m,n,k,k,r} = 0$ ensures that no outgoing transmission of message M_m occurs from E_k to any other node.

Furthermore, if no physical link exists between two nodes in the network (i.e., $l_{r,q} = 0$), message transmission between these nodes is not permitted. This restriction is captured by Eq. 5.4.

It is also possible for the source and destination tasks of a message to be executed on the same processing element. In this case, the variable $y_{m,k,k,k,k}$ is set to 1, indicating that the message is transmitted internally without traversing the network. Consequently, the message should not be assigned to any other links, which is enforced by ensuring that $y_{m,k,k,n,q} = 0$ for all $(n, q) \neq (k, k)$. This scenario is captured by the constraints defined in Eq. 5.5 and Eq. 5.6.

The Eq. 5.7 is introduced to prevent routing cycles between two nodes, which would result in invalid message forwarding. A cycle occurs when a message visits the same node more than once. This constraint ensures that a message can either enter or exit a node, but not both, thereby avoiding revisits.

The variable $y_{m,k,n,r,q}$ should be set to 0 for all values of k except for the E_k on which the source task of message M_m is mapped. Similarly, $y_{m,n,k,r,q}$ should be set to 0 for all values of k except for the E_k on which the destination task of message M_m is mapped. These constraints are enforced using Eq. 5.8 and Eq. 5.9.

The approach presented here utilizes single-path routing, where each message is forwarded through a unique path to its designated destination node. In this

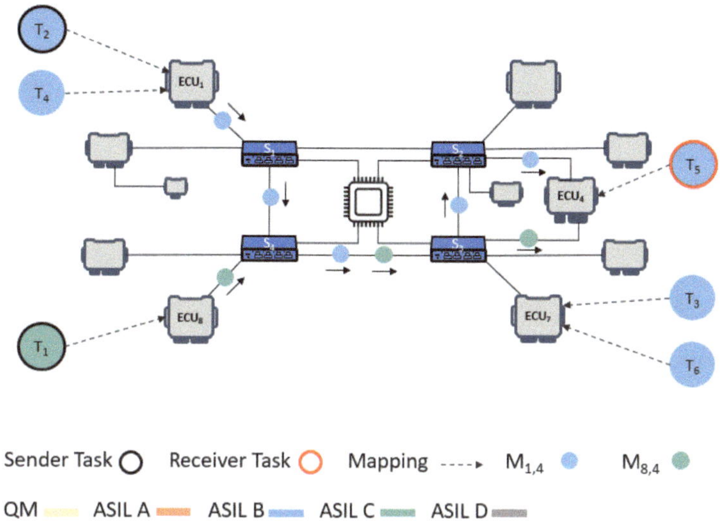

Figure 5.1: A vehicle architecture consisting of control and intermediate nodes with application tasks mapped to ECUs. Single paths are generated to transmit communication messages (represented by colored dots) from sender nodes to corresponding receiver node

routing model, a message is not duplicated or branched; instead, it traverses the network via a single sequence of intermediate nodes. To ensure this behavior, each message is allowed to depart from a given node to at most one next-hop node, and similarly, it can arrive at a given node from at most one previous node. These constraints are enforced through Eq. 5.10 and Eq. 5.11. In cases where redundant routing is required for fault tolerance or other purposes, these constraints can be relaxed to allow multiple paths for message delivery.

Figure 5.1 provides an example of unicast routing. The routing is illustrated for two communication messages: $M_{1,4}$, which is sent from ECU_1 to ECU_4 (with source task T_2 and destination task T_5), and $M_{8,4}$, which is sent from ECU_8 to ECU_4 (with source task T_1 and destination task T_5).

To enforce at least one outgoing transmission from the source node and at least one incoming reception at the destination node when the source and destination tasks of a message are mapped to different ECUs, following constraints are used:

$$\sum_{\substack{N_n \in N \\ N_r \in N \\ h \in [1..4]}} y_{m,k,n,k,r} \cdot x_{j,n,h} \geq \sum_{h_1=1}^{4} x_{i,k,h_1} - \sum_{h_2=1}^{4} x_{j,k,h_2} + \sum_{\substack{h_1=1 \\ h_2=1}}^{4} x_{i,k,h_1} \cdot x_{j,k,h_2}$$

Eq. 5.12

$$\sum_{\substack{N_n \in N \\ N_r \in N \\ h \in [1..4]}} y_{m,n,k,r,k} \cdot x_{i,n,h} \geq \sum_{h_1=1}^{4} x_{j,k,h_1} - \sum_{h_2=1}^{4} x_{i,k,h_2} + \sum_{\substack{h_1=1 \\ h_2=1}}^{4} x_{j,k,h_1} \cdot x_{i,k,h_2}$$

Eq. 5.13

Eq. 5.12 ensures that if the source task T_i is mapped to ECU E_k, and the destination task T_j is mapped to a different ECU, then at least one routing variable $y_{m,k,n,k,r}$ must be set to 1, indicating that the message M_m is transmitted from node E_k to some other node. Similarly, constraint Eq. 5.13 ensures that if the destination task is mapped to ECU E_k, and the source is located on a different ECU, then at least one variable $y_{m,n,k,r,k}$ must be set to 1, indicating that the message arrives at node E_k from another node. If route redundancy is required, the desired number of redundant paths must be added to the right-hand side of each of the above constraints.

The above constraints are applicable when the source and destination tasks of a message are not subject to decomposition. However, if ASIL decomposition is applied to either the source or destination tasks, or both, then additional considerations are required. Specifically, the number of outgoing transmissions from the source node must be at least equal to the number of decomposed destination tasks, and similarly, the number of incoming transmissions to the destination node must be at least equal to the number of decomposed source tasks. These requirements are enforced through constraints Eq. 5.14 and Eq. 5.15, respectively.

if $T_j \in T'$:

$$\sum_{\substack{N_n \in N \\ N_r \in N \\ h \in [1..4]}} y_{m,k,n,k,r} \cdot x_{j,n,h} \geq \sum_{\substack{h_1=1 \\ h_2=1}}^{4} x_{i,k,h_1} \cdot \alpha_{j,h_2} - \sum_{h_2=1}^{4} x_{j,k,h_2} + \sum_{\substack{h_1=1 \\ h_2=1}}^{4} x_{i,k,h_1} \cdot x_{j,k,h_2}$$

Eq. 5.14

if $T_i \in T'$:

$$\sum_{\substack{N_n \in N \\ N_r \in N \\ h \in [1..4]}} y_{m,n,k,r,k} \cdot x_{i,n,h} \geq \sum_{\substack{h_1=1 \\ h_2=1}}^{4} x_{j,k,h_1} \cdot \alpha_{i,h_2} - \sum_{h_2=1}^{4} x_{i,k,h_2} + \sum_{\substack{h_1=1 \\ h_2=1}}^{4} x_{j,k,h_1} \cdot x_{i,k,h_2}$$

Eq. 5.15

According to Eq. 5.16, intermediate nodes, which are neither source nor destination nodes, must have an equal number of incoming and outgoing communication messages.

$$\sum_{N_q \in N} y_{m,n,r,k,q} \cdot \left(1 - \sum_{h=1}^{4} x_{i,k,h}\right) = \sum_{N_q \in N} y_{m,n,r,q,k} \cdot \left(1 - \sum_{h=1}^{4} x_{j,k,h}\right)$$

Eq. 5.16

To ensure freedom from interference, as outlined by ISO 26262, when ASIL decomposition is applied, it is crucial to ensure that messages originating from decomposed source nodes or destined for decomposed destination nodes are routed over distinct links. This condition is guaranteed by equations Eq. 5.17 and Eq. 5.18. The variable $rl_{n,q}$ represents the number of physical links between two nodes and is provided as input. For instance, if the nodes are connected via two separate Ethernet cables, then $rl_{n,q} = 2$.

If $T_i \in T'$ **and** $n \neq q$:

$$\sum_{N_r \in N} \left(y_{m,r,k,n,q} + y_{m,r,k,q,n}\right) \leq rl_{n,q}$$

Eq. 5.17

If $T_j \in T'$ **and** $n \neq q$:

$$\sum_{N_r \in N} \left(y_{m,k,r,n,q} + y_{m,k,r,q,n}\right) \leq rl_{n,q} \qquad \text{Eq. 5.18}$$

The routings resulting from the above constraints are not necessarily disjoint, meaning they may share common nodes, but there should be no shared links between senders and receivers. To ensure that the routings are fully disjoint, with no shared nodes or links, the following constraints are applied:

$$\sum_{N_q, N_n \in N} y_{m,k,n,r,q} \leq 1, \quad r \neq k \qquad \text{Eq. 5.19}$$

$$\sum_{N_r, N_n \in N} y_{m,n,k,r,q} \leq 1, \quad q \neq k \qquad \text{Eq. 5.20}$$

Figures 5.2(a) and 5.2(b) provide an example of message routing under the FFI constraint and the combination of FFI and disjoint path constraints, respectively. In this example, task T_1 is decomposed into two redundant subtasks, mapped to ECU_1 and ECU_8. The routing of messages from T_1 to T_5, which is executed on ECU_4, is illustrated. As shown in Figure 5.2(a), applying only the FFI constraint may still result in shared intermediate nodes (e.g., switch S_3). It is worth noting that, in this case, switch S_3 must be developed in compliance with ASIL D. In contrast, Figure 5.2(b) demonstrates that incorporating the disjoint path constraint ensures completely independent routes with no shared nodes.

To prevent network link overload, Eq. 5.21 ensures that the total transmission load assigned to each physical link does not exceed its available bandwidth. In this formulation, the message size is given in bits, the period is in seconds, and the link speed (i.e., bandwidth) is specified in bits per second. The parameter $br_{r,q}$ denotes the bandwidth of the physical link $l_{r,q}$. The constraint is formulated as follows:

$$\sum_{M_m \in M} \left(\frac{M_{m,s}}{M_{m,p}} \cdot y_{m,k,n,r,q}\right) \leq br_{r,q}, \quad r \neq q \qquad \text{Eq. 5.21}$$

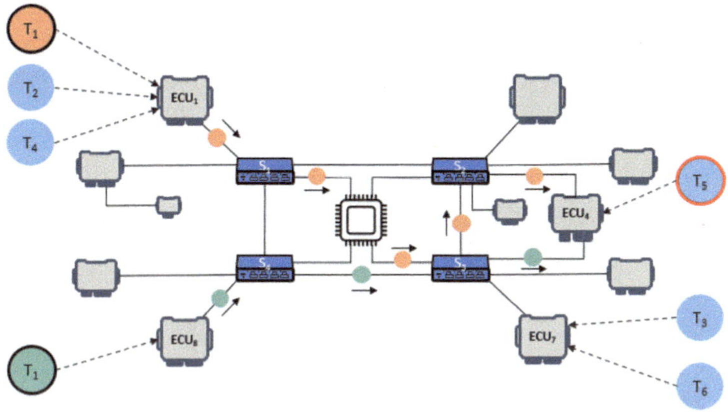

(a) Routing with FFI constraint only — shared node S_3 is used by both paths

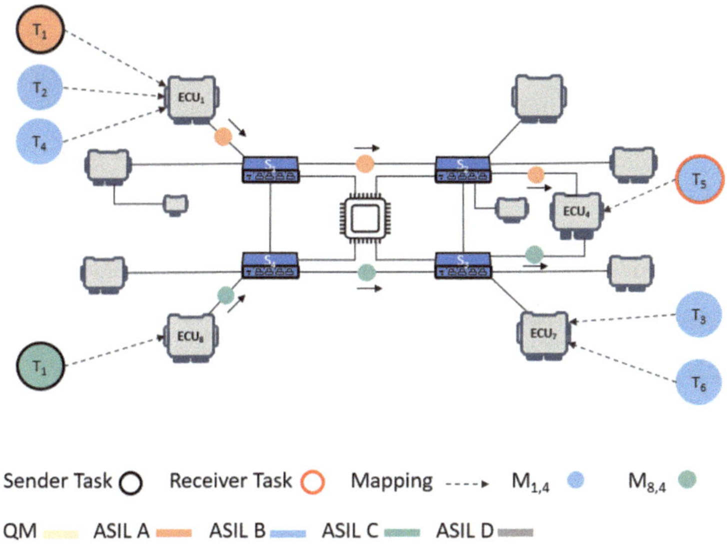

(b) Routing with both FFI and disjoint path constraints — paths are fully disjoint

Figure 5.2: Comparison of routing strategies in a vehicle network consisting of control and intermediate nodes with application tasks mapped to ECUs

5.2.2 Message Scheduling

To enable time-triggered scheduling of communication messages transmitted over network links, the following set of constraints is employed:

$\forall T_i, T_j \in T; T_i \neq T_j; \forall N_k, N_n, N_r, N_q \in N; \forall M_m \in M; M_m^{src} = T_i; M_m^{dst} = T_j$:

$$\sigma_{m,k} \cdot x_{j,k,h} \leq M_{m,p}, \quad \forall h \in [1 \ldots 4] \qquad \text{Eq. 5.22}$$

$$\sigma_{m,q} \geq y_{m,k,n,r,q} \cdot \left(\sigma_{m,r} + \frac{M_{m,s}}{br_{r,q}}\right), \quad r \neq q, l_{r,q} = 1 \qquad \text{Eq. 5.23}$$

The arrival time of each message M_m on node N_k is denoted by $\sigma_{m,k}$. Eq. 5.22 ensures that each message M_m is schedulable within its specified period $M_{m,p}$. As in the previous section, the destination task of message M_m is T_j. Therefore, if $x_{j,k,h} = 1$, it indicates that the destination task T_j is mapped to node N_k, which is thus the destination node for message M_m. This requires that the arrival time of the message at this node, must not exceed its deadline, which in this context is assumed to be equal to the message period $M_{m,p}$.

To guarantee proper timing consistency across the communication path, also known as path dependency, Eq. 5.23 is used. It enforces that the release time of a message at the receiving node is no earlier than its release time at the sending node plus the transmission delay over the corresponding communication link. To avoid division by zero, the bandwidth variable $br_{r,q}$ is assigned a value of 1 in cases where no physical link exists between nodes N_r and N_q.

The overlapping-free condition for two messages M_m and M_u routed over the same node N_k is ensured using the following constraints:

$\forall M_m, M_u \in M;\ M_m \neq M_u;\ \forall N_k, N_n, N_r \in N;$

$$\forall w_m \in \left[0, \frac{\text{LCM}(M_{m,p}, M_{u,p})}{M_{m,p}} - 1\right]; \quad \forall w_u \in \left[0, \frac{\text{LCM}(M_{m,p}, M_{u,p})}{M_{u,p}} - 1\right]:$$

$$\sigma_{m,k} + M_{m,p} \cdot w_m + \sum_{N_q \in N} y_{m,n,r,k,q} \cdot l_{k,q} \cdot \frac{M_{m,s}}{br_{k,q}} \leq \sigma_{u,k} + M_{u,p} \cdot w_u$$

or

$$\sigma_{u,k} + M_{u,p} \cdot w_u + \sum_{N_q \in N} y_{u,n,r,k,q} \cdot l_{k,q} \cdot \frac{M_{u,s}}{br_{k,q}} \leq \sigma_{m,k} + M_{m,p} \cdot w_m$$

Eq. 5.24

The above constraint represents an either-or condition, which must be reformulated into a linear form. Similar to the approach discussed in Section 4.2.1, the Big-M method is employed by introducing a binary decision variable η_{m,u,k,w_m,w_u}. The linearized form of this constraint is captured by Eq. 5.25 and Eq. 5.26, where M denotes a sufficiently large positive constant.

$$\sigma_{u,k} + M_{u,p} \cdot w_u - \sigma_{m,k} - M_{m,p} \cdot w_m - \sum_{N_q \in N} y_{m,n,r,k,q} \cdot l_{k,q} \cdot \frac{M_{m,s}}{br_{k,q}} \geq -M \cdot (1 - \eta)$$

Eq. 5.25

$$\sigma_{m,k} + M_{m,p} \cdot w_m - \sigma_{u,k} - M_{u,p} \cdot w_u - \sum_{N_q \in N} y_{u,n,r,k,q} \cdot l_{k,q} \cdot \frac{M_{u,s}}{br_{k,q}} \geq -M \cdot \eta$$

Eq. 5.26

Message dependency, referring to the interdependence and ordering of message transmissions over the network, plays a crucial role in ensuring the correct operation of distributed systems. This concept is particularly important because certain messages may rely on the reception or processing of others before they can be correctly transmitted or acted upon [140].

Previously, functional dependencies between tasks were introduced using the binary variable $d_{i,j}$, which indicates that task T_i must be executed before task T_j when $d_{i,j} = 1$. Since messages are associated with the execution of these tasks, the same dependency pattern extends naturally to the messages exchanged between them.

To capture this dependency, the following constraints are employed:

$\forall M_m \in M; \forall T_i, T_j \in T; \forall E_k, E_r \in E; M_m^{src} = T_i; M_m^{dst} = T_j$:

$$\sigma_{m,k} \geq \sum_{h=1}^{4} \left(\tau_{i,k} + wcet_{i,k,h}\right) \cdot x_{i,k,h} \qquad \text{Eq. 5.27}$$

$$\tau_{j,k} \geq \sum_{N_n \in N} \left(\sigma_{m,n} + \frac{M_{m,s}}{br_{n,k}}\right) \cdot l_{n,k} \cdot y_{m,r,k,n,k} \qquad \text{Eq. 5.28}$$

Eq. 5.27 ensures that on the source node of a message, the corresponding task must first be executed before the related message can be released into the network. Similary but this time for destination node, Eq. 5.28 guarantees that the task on the destination node will only start execution after the corresponding message has been completely received.

In addition to these constraints, a slight modification is required in the formulation of task dependencies. In Section 4.2.1, we assumed a precomputed worst-case communication time between two dependent tasks $wcrt_{i,j}$. However, since the communication time is now explicitly modeled based on network parameters, the fixed delay-based formulation (Eq. 4.15) should be removed, and the dependency constraint (Eq. 4.14) should be revised as:

$\forall(T_i, T_j) \in T; (E_k, E_m) \in E; T_i \neq T_j; d_{i,j} = 1:$

$$\tau_{i,k} + \sum_{\substack{h_1 \in [1...4] \\ h_2 \in [1...4]}} x_{i,k,h_1} \cdot wcet_{i,k,h_1} \cdot x_{j,m,h_2} \leq \tau_{j,m} \qquad \text{Eq. 5.29}$$

The objective function is defined to minimize the overall latency. The formulas introduced in the previous chapter are also applicable in this context.

5.3 Experimental Results

The effectiveness of the overall framework is evaluated using the motivational example presented in Section 4.1, applied to a network consisting of four ECUs and three switches (see Figure 5.3). Seven messages are considered, with sender and destination nodes aligned with the task dependencies. Each message is assigned a period of 100 ms and a payload size of 100 bytes. The ECUs

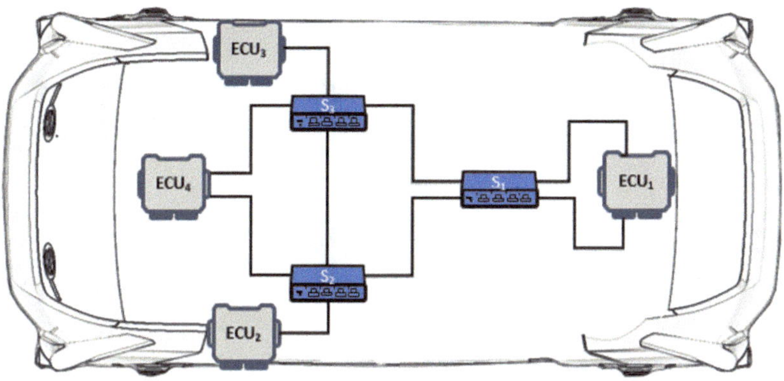

Figure 5.3: An example vehicle architecture consisting of four ECUs and three switches

and switches are interconnected using 100 Mb/s Ethernet links. The ECU specifications are similar to those presented in Table 4.1.

The results are compared under three different assumptions:

1. **Scenario A**: All tasks are classified as ASIL B, eliminating the need for decomposition. Message routing is performed using a basic single-path strategy.

2. **Scenario B**: Tasks T_1 and T_6 are assigned ASIL D. Routing is performed using a simple single-path strategy, without any additional safety constraints.

3. **Scenario C**: Tasks T_1 and T_6 are assigned ASIL D. Routing is constrained solely by the FFI requirement.

4. **Scenario D**: Tasks T_1 and T_6 are assigned ASIL D. Routing is constrained by both the FFI requirement and disjoint path constraints to enhance communication safety and fault isolation.

Figure 5.4 shows the minimum latency achieved under each scenario. As expected, the latency generally increases from Scenario A to Scenario D. Figure 5.5 provides a comparative analysis of the solving time and model generation time across all four scenarios.

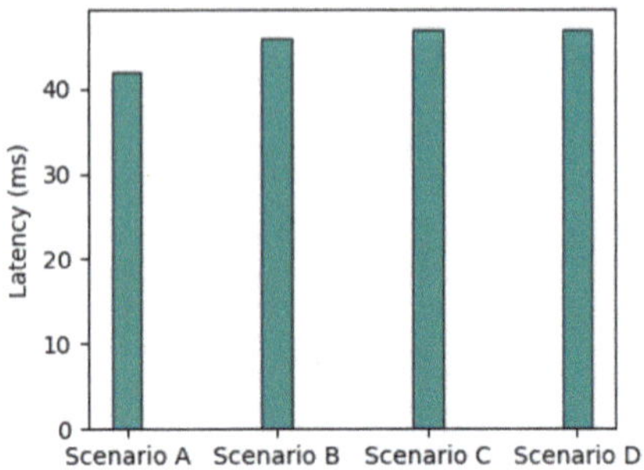

Figure 5.4: Minimum latency achieved under different scenarios

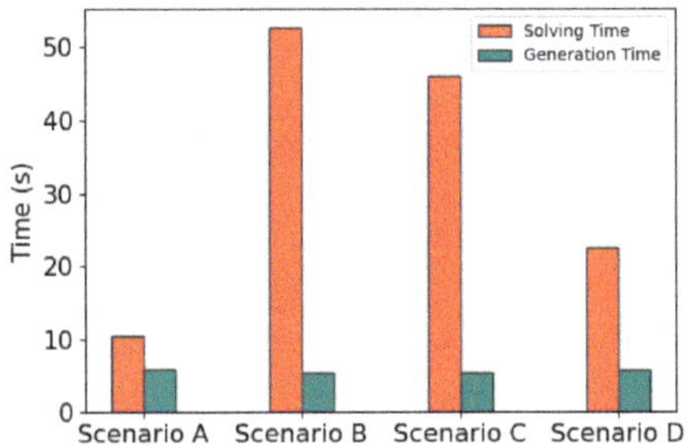

Figure 5.5: Comparison of solving time and model generation time for different scenarios

6 Evaluation

This chapter presents a detailed assessment of the proposed framework, focusing on its performance, scalability, and applicability in automotive system design. The evaluation is conducted through a series of structured experiments, and the effectiveness of the approach is investigated using diverse case studies and real-world benchmarks, enabling a critical examination of the framework's theoretical and practical aspects. All experiments were conducted on a PC equipped with an Intel Core i7-8565U CPU running at 1.8 GHz and 32 GB of RAM. The optimization models were implemented in Python and solved using the Gurobi optimizer.

Given that the proposed framework incorporates multiple constraints and jointly addresses two interdependent subproblems—criticality-aware resource allocation and message routing—the evaluation is structured to assess each component both independently and as part of the integrated solution. Accordingly, the assessment is organized into three categories: (i) evaluation of criticality-aware resource allocation, (ii) evaluation of message routing, and (iii) evaluation of the complete framework through unified single-step optimization.

Furthermore, the chapter examines the framework's behavior under varying safety constraints and communication topologies, providing a comprehensive analysis of its robustness and adaptability. The results aim to substantiate the effectiveness of the proposed approach and highlight its potential for integration into real-world automotive applications.

6.1 Evaluation of Criticality-Aware Resource Allocation

This experiment investigates the synthesis time associated with optimizing resource allocation in mixed-criticality automotive systems. The primary objective is to evaluate the impact of mapping, task scheduling, and ASIL decom-

D. Mohammad Zaheri, *Safety-Oriented Optimization of Automotive E/E Architectures*, Wissenschaftliche Reihe Fahrzeugtechnik Universität Stuttgart,
https://doi.org/10.1007/978-3-658-53045-7_6

position constraints—introduced in Chapter 4—on the computational solving time of the proposed MIP-based framework. The study is conducted across seven distinct scenarios, each focusing on different system attributes, such as the number of ECUs and the number of tasks.

In all scenarios, the main optimization goal is to minimize the total cost. Additionally, some scenarios incorporate a multi-objective optimization strategy, where latency is considered as a secondary objective alongside cost. This setup enables the analysis of the influence of additional constraints on solving time.

The approach considered follows the Mixed-ASIL strategy. The task communication graph is randomly generated using the Erdős–Rényi model [141], with a connectivity probability of 0.9. ECU failure rates are randomly assigned within the range of 8×10^{-7} to $4 \times 10^{-6}[h^{-1}]$. Cost values are assigned randomly according to the *experiential-I* cost model (see Table 3.1), which, as shown in Section 4.3.3, results in the highest solving time among all evaluated formulations. This model is used throughout the experiments to represent the worst-case scenario and to evaluate the scalability of the proposed framework under the most computationally demanding conditions. Furthermore, WCET values are randomly sampled between 1 ms and 20 ms, while WCRT values range from 1 ms to 15 ms.

First Case Study

In the initial case study, a hardware architecture with four ECUs is considered. All tasks are ASIL B, and all ECUs support up to ASIL C, eliminating the need for ASIL decomposition. Tasks are non-periodic, with simplified scheduling constraints as described in [116], and no localization or memory constraints are applied. This experiment evaluates solving time as the number of tasks increases from 10 to 70. Figure 6.1(a) shows results for multi-objective optimization, while Figure 6.1(b) shows the outcomes when only development cost is optimized. As illustrated, the multi-objective optimization requires more time to solve. Furthermore, a noticeable exponential increase in solving time is observed, especially as the number of tasks surpasses thirty.

Second Case Study

The second experiment evaluates how the number of ECUs affects solving time. An application with 10 non-periodic tasks is considered, with dependencies generated via an Erdős–Rényi model (connectivity 0.9). No ASIL decomposition is applied. The number of ECUs ranges from 10 to 50. Results for both multi-objective optimization (development cost and latency; Figure 6.1(c)) and single-objective optimization (development cost; Figure 6.1(d)) show that solving time increases with the number of ECUs.

Third Case Study

To further assess the impact of system complexity on the solving time of the proposed framework, an experiment is conducted in which the system initially comprises 10 ECUs and 10 tasks. The system dimensions are then scaled up to encompass 50 ECUs and 50 tasks. As in previous case studies, tasks are assumed to be non-periodic, no ASIL decomposition is required, and the objective is to minimize the overall development cost. Figure 6.1(e) illustrates the observed growth in both solving and generation times as the system complexity increases.

Fourth Case Study

Deriving schedules for periodic tasks becomes significantly more intricate when task periods are not harmonic, due to the complexity introduced by least common multiple and hyperperiod calculations. To illustrate this, an experiment was conducted to evaluate the impact of non-harmonic periods on scheduling complexity. The case study considers a hardware architecture comprising 10 ECUs, with the number of tasks varied from 3 to 15. The objective is to minimize the overall system cost, under the assumptions that there is no ASIL decomposition and no memory or localization constraints. Additionally, a constraint on CPU utilization is imposed to ensure feasible execution. One-third of the tasks are assigned a period of 50 ms, another third 60 ms, and the remaining third 80 ms, resulting in a hyperperiod of 1200 ms. As shown in Figure 6.1(f), both the solving time and generation time increase significantly as the problem size grows. This is primarily due to the added complexity introduced by non-harmonic periods and the larger hyperperiod.

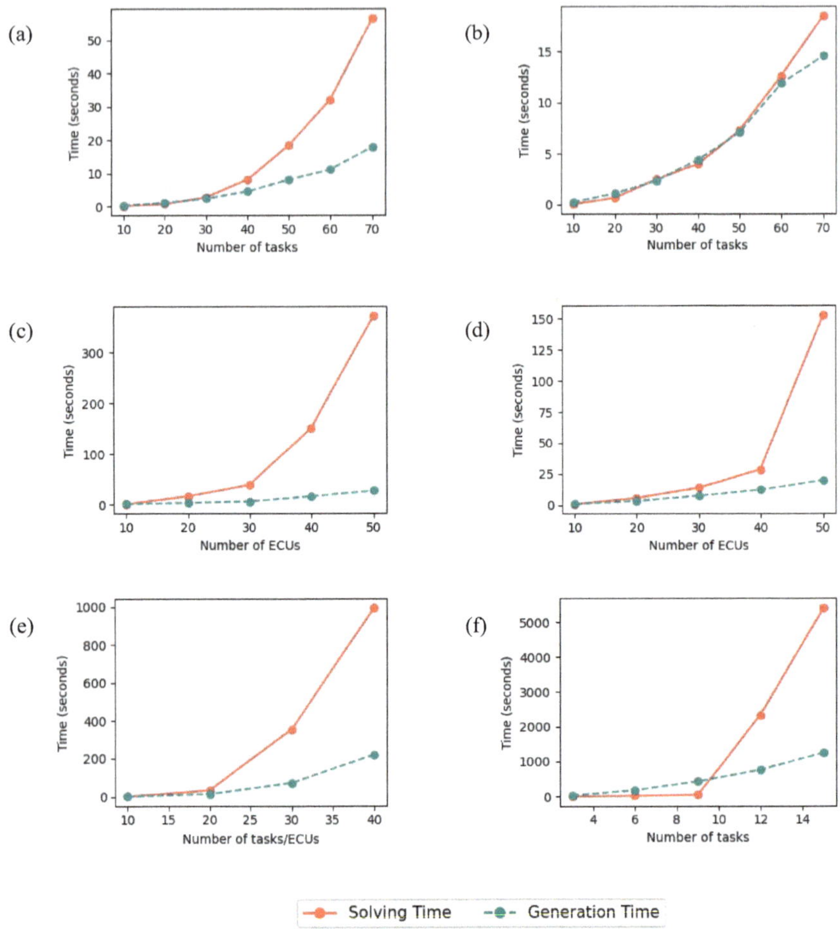

Figure 6.1: Performance evaluation of the proposed framework. In all use cases, the minimization of overall development cost is set as the primary objective. (a) Multi-objective optimization is applied to a topology comprising 4 ECUs with ASIL C and tasks with ASIL B. (b) Single-objective optimization is applied to the same topology as in (a). (c) Multi-objective optimization is applied to an application consisting of 10 tasks with ASIL B. (d) Single-objective optimization is applied to the same application as in (c). (e) All ECUs have ASIL C, and all tasks have ASIL B. (f) Single-objective optimization is applied to a topology comprising 10 ECUs and periodic tasks.

Fifth Case Study

This case study evaluates the impact of ASIL decomposition on the generation and solving time of the proposed framework. We consider a network topology consisting of 10 ECUs, where all tasks are non-periodic, assigned ASIL D, and subject to ASIL decomposition. To explore the full flexibility of decomposition strategies, we relax the constraint on ASIL compatibility between tasks and ECUs, allowing the solver to select any valid decomposition schemas that minimizes development cost. This increases computational complexity, as the model must evaluate all possible decomposition combinations. The number of tasks is varied from 10 to 50, and the resulting solving times are presented in Figure 6.2.

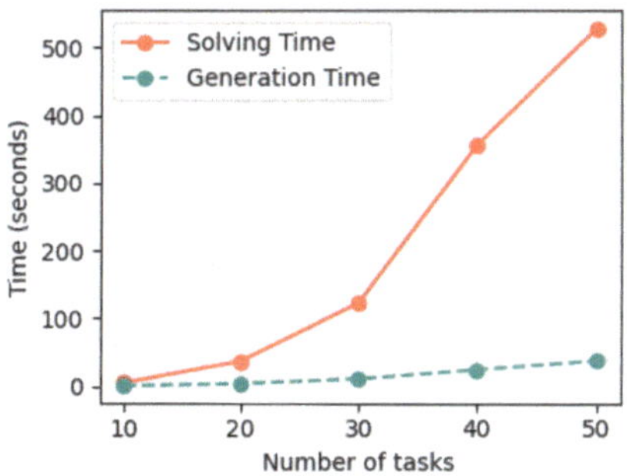

Figure 6.2: Evaluation of solving and generation times for the proposed framework under ASIL decomposition. All tasks are assigned ASIL D, with no restrictions on the decomposition combinations. The system comprises 10 ECUs, and the optimization objective is to minimize development cost.

Sixth Case Study

To investigate the impact of ASIL decomposition on solving time under decomposition schema restrictions, an experiment was conducted. The results are shown in Figure 6.3. In this case study, the network topology consists of 4 ECUs, and the objective is to minimize overall system cost. In the first experiment, all tasks were assigned ASIL D, while the ECUs supported up to ASIL C.

This scenario required more computation time, since all ASIL decomposition schemas—except for ASIL D + QM—needed to be tested by the algorithm. In the second experiment, all tasks had ASIL C, and the ECUs supported up to ASIL B. As shown in Figure 6.3, the computation time in this scenario was significantly lower than in the first, as expected.

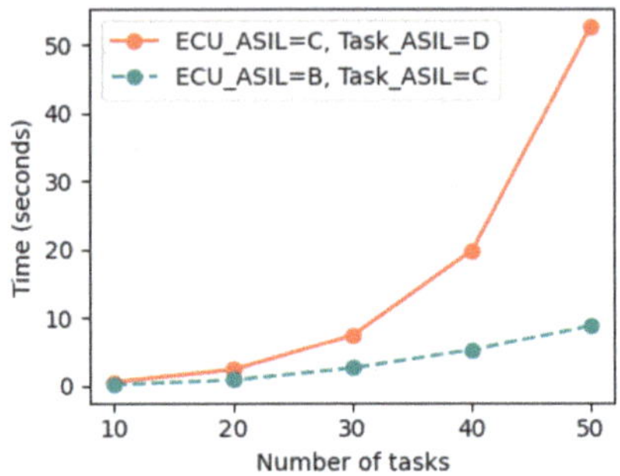

Figure 6.3: Solving time evaluation of the proposed framework under ASIL decomposition, considering a network topology of 4 ECUs. The objective is overall cost minimization, with different ASIL assignments for tasks and ECUs.

Seventh Case Study

An important factor influencing the solving time of the proposed framework is the assigned development cost values. The framework was applied to different cost models, as presented in Table 3.1.

In the first scenario, a topology consisting of four ECUs with non-periodic tasks, all assigned ASIL B is considered. ASIL decomposition is not applied in this case. The optimization problem is solved using a multi-objective approach, where the primary objective is to minimize the overall development cost and the secondary objective is to minimize the overall latency. This setup is evaluated under various cost models. The number of tasks is increased from 10 to 70, and the corresponding results are illustrated in Figure 6.4(a).

In the second scenario, the same topology and multi-objective formulation are retained, but all tasks are assigned ASIL D, making them subject to ASIL decomposition. To increase complexity, the ASIL compatibility constraint is relaxed, allowing any decomposition schema to be selected. The results of this experiment are illustrated in Figure 6.4(b). According to the findings, in the scenario where ASIL decomposition is not applied, there is no consistent pattern in the solving times across different cost models. However, when ASIL decomposition is employed, the logarithmic cost model consistently results in the shortest solving time

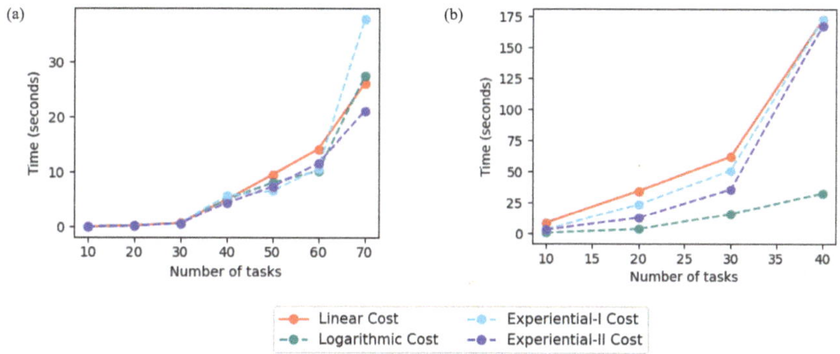

Figure 6.4: Solving time comparison under different development cost models for increasing task counts on a topology with 4 ECUs. (a) Tasks with ASIL B and no decomposition. (b) Tasks with ASIL D with decomposition and relaxed ASIL compatibility constraints.

6.1.1 Comparison with Genetic Algorithm-Based Approach

In this case study, the proposed ILP-based framework is compared with the GA approach introduced in Section 4.2.3. The results of applying both methods to randomly generated task graphs mapped to a topology with 4 ECUs are illustrated in Figure 6.5. It is noteworthy that task scheduling constraints are not considered in this experiment. All tasks are assumed to have ASIL D, and the objective is to find the optimal allocation that minimizes the overall development cost. As expected, the ILP method consistently identifies the exact minimum due to its deterministic optimization nature, whereas the GA yields approximate solutions. However, a direct comparison of computation times is not meaningful, as the ILP executes a single optimization run, while the GA's solving time depends heavily on the number of generations and other stochastic factors.

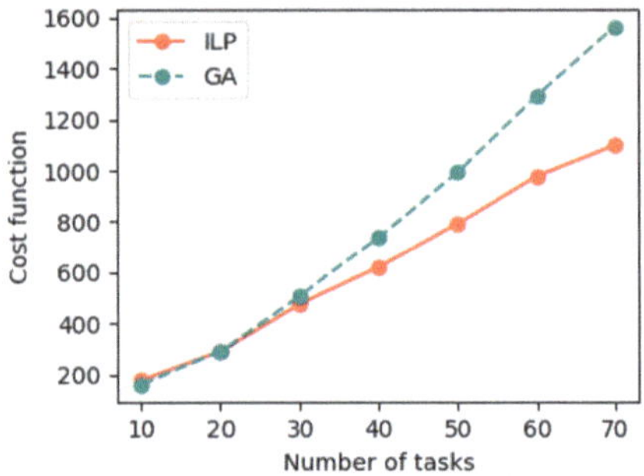

Figure 6.5: Comparison of cost function results between the proposed ILP model and GA method

6.2 Evaluation of Automated Message Routing

This experiment evaluates the solving time associated with generating message routing constraints. A randomly generated network topology based on the Erdős–Rényi graph model with a connectivity probability of 0.95 is used. Message periods are randomly selected in the range of 20 to 100 ms. All link speeds are assumed to be 100 Mbit/s, and message sizes vary between 64 bytes and 1500 bytes.

Since routing and task mapping to ECUs are coupled in the formulation, tasks are assumed to be pre-mapped to ECUs to isolate the solving time related specifically to routing constraints. The ECU assignments for each task are performed randomly to ensure varied distribution across the network.

In the first case study, no ASIL decomposition is applied, resulting in routing based on a single routing scheme. The model is applied to a topology consisting of 10 ECUs and 10 tasks. For each message, a random source task and destination task are selected, ensuring that the source and destination tasks are distinct.

A link utilization constraint is incorporated to ensure that the total data transmitted over any link does not exceed its capacity. The number of messages is incrementally increased from 10 to 50, and the corresponding solving times are recorded. The results are illustrated in Figure 6.6(a), which shows that, as expected, the solving time increases with the number of messages. Figure 6.6(b) further analyzes the impact of increasing the number of nodes on the solving time for a topology with 10 tasks and 10 messages.

The second case study assumes all tasks undergo ASIL decomposition using a uniform ASIL B + ASIL B strategy. The model is applied to a network topology with 10 ECUs and 10 tasks. To prevent infeasible solutions due to stringent communication limits, the link utilization constraint is relaxed in this scenario.

Solving times are evaluated and compared across three configurations: (i) when only single-path routing constraints are enabled, (ii) when FFI constraints are also activated, and (iii) when both disjoint routing and FFI constraints are enforced simultaneously. The results of this comparison are illustrated in Figure 6.6(c).

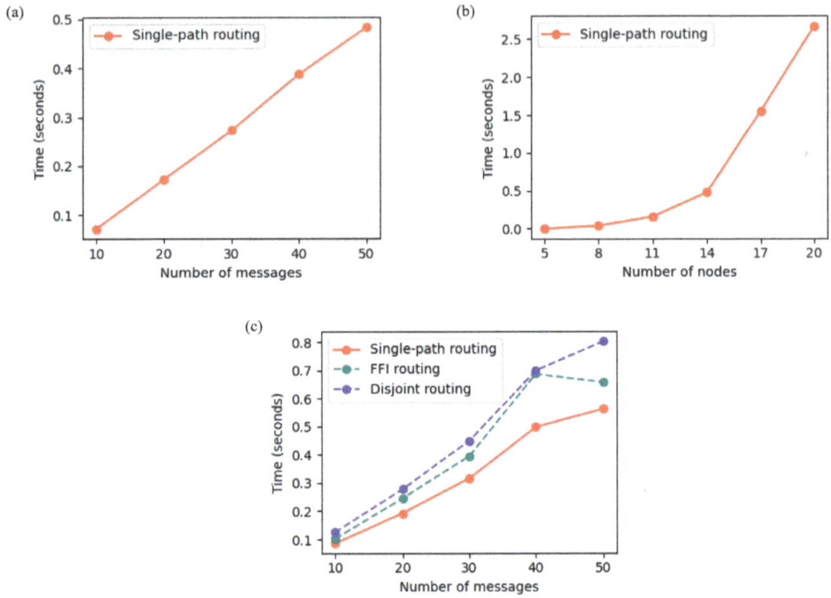

Figure 6.6: Evaluation of message routing solving time under different scenarios using topologies randomly generated based on the Erdős–Rényi graph model. (a) Impact of increasing the number of messages on solving time, using topologies consisting of 10 ECUs and 10 tasks. (b) Impact of increasing the number of nodes on solving time, with a fixed number of 10 tasks and 10 messages. (c) Comparison of solving time under different constraint configurations—single-path only, single-path with FFI, and single-path with both FFI and disjoint routing—assuming ASIL B + B decomposition, using topologies with 10 ECUs and 10 tasks.

6.3 Evaluation of the Complete Framework

This experiment evaluates the computational performance of the proposed framework, focusing on its solving time. To this end, we consider the UNICARagil zonal topology comprising 11 ECUs and 4 switches (see Figure 3.1b), which serves as the hardware architecture for the evaluation. The objective is to minimize the overall development cost while satisfying all model constraints, except for memory and localization, which are disabled in this scenario.

Task and message periods are assumed to be even, and a fixed number of 10 messages are generated between randomly selected source and destination tasks. The number of tasks is increased from 10 to 50, and for each configuration, both the model generation and solving times are recorded. All tasks are assumed to be of ASIL B, and the ECUs are capable of handling tasks up to ASIL B as well.

As illustrated in Figure 6.7, the solving time increases with the number of tasks, reflecting the growing complexity of the optimization problem.

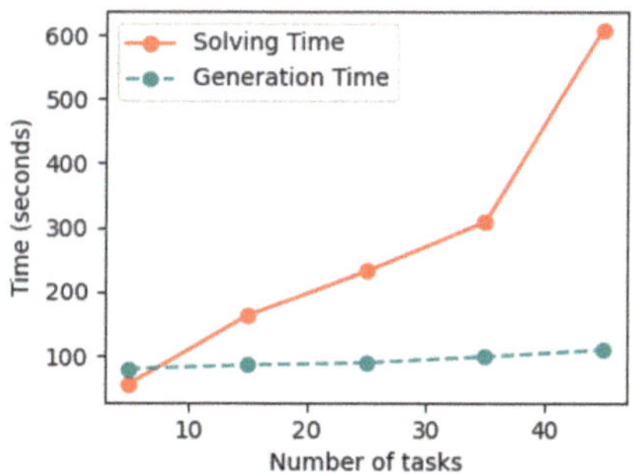

Figure 6.7: Solving time evaluation of the overall framework. The UNICARagil zonal architecture is selected as the hardware topology. The objective is to minimize the overall development cost while satisfying all model constraints, except for memory and localization.

6.4 Results Analysis and Strategies for Solving Time Improvement

Based on the results obtained from various experiments, the proposed method demonstrates practicality for industrial applications. While the solving time may be high in certain cases, it remains manageable, especially considering the complexity of the problem at hand. The results indicate that the method can be a valuable tool for real-world applications, offering feasible solutions within a reasonable time frame. To further improve solution efficiency, the following strategies are proposed:

First, it may be beneficial to approach the problem in two parts: resource allocation and message scheduling and routing. Solving these components separately, possibly using an iterative approach, could help reduce the complexity of the problem and lead to faster solutions.

Second, techniques such as adjusting the `MIPGap` parameter can be effective when a near-optimal solution is acceptable. By setting `model.setParam(GRB.Param.MIPGap, 0.2)`, the solver is permitted to terminate once a solution within 20% of the optimal bound is found. This relaxation can reduce the solving time while still yielding a practically useful result. This adjustment is applied specifically to the fourth case study, discussed in Section 6.1 and illustrated in Figure 6.1(f), with the only difference being that the number of ECUs is set to eight in this instance. The impact of this adjustment is illustrated in Figure 6.8. Figure 6.8(a) compares the solving times of the standard and MIP-gap-based formulations, while Figure 6.8(b) compares the corresponding objective values, representing the total cost. With a 20% MIP gap, we observed an approximate 25% reduction in solving time, with only a minor trade-off in the optimality of the solution.

Another promising approach is the use of a binary search algorithm, which can be applied in certain scenarios to further reduce solving time. It begins by solving the optimization model without a cost bound to obtain an initial feasible solution. The objective value of this solution is assigned as the initial upper bound, representing the best known feasible cost. A lower bound is then estimated—often by computing the minimum possible cost.

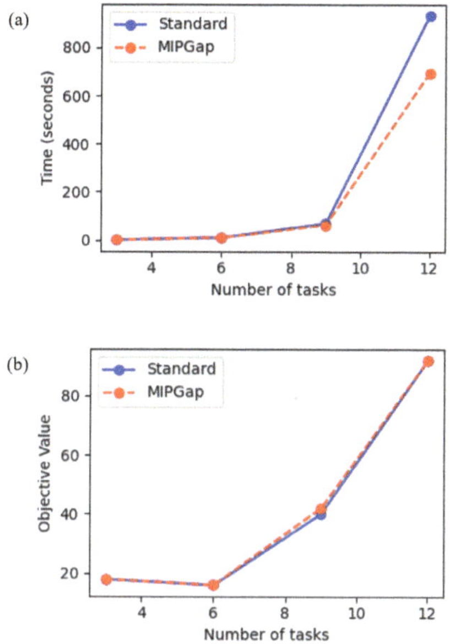

Figure 6.8: Comparison of solving time and objective value in the modified fourth case study (8 ECUs) using a `MIPGap` of 0.2: (a) Solving time reduction, (b) Objective value (cost).

The algorithm then enters an iterative refinement loop: at each iteration, a new cost limit is set to the midpoint between the current lower and upper bounds. The model is re-solved with this new cost constraint. If a feasible solution is found, the upper bound is updated to this lower cost, thereby tightening the range. If the model is infeasible, the lower bound is raised to the midpoint, excluding non-viable solutions. This process continues until the difference between the upper and lower bounds is less than a defined tolerance (ϵ), ensuring convergence to the minimum feasible cost within a controlled precision. The pseudo-code for this algorithm is presented in Algorithm 6.2.

Finally, to reduce model generation time, it can be beneficial to simplify the problem formulation by minimizing the number of decision variables. Variables

Algorithm 6.2 Binary Search Optimization for Cost

Require: *low* (initial lower bound), *high* (initial upper bound), ϵ (tolerance)
Ensure: *best_cost*

```
best_cost ← high
while high − low > ϵ do
    mid ← (low+high)/2
    Solve the model with cost limit = mid
    if model is feasible then
        best_cost ← mid
        high ← mid                          ▷ Update upper bound
    else
        low ← mid                           ▷ Update lower bound
    end if
end while
return best_cost
```

such as $y_{m,k,n,r,q}$, which represent message routing across multiple nodes and links, contribute substantially to the model's size and complexity. In scenarios where ensuring freedom from interference is essential, a simplification strategy may involve omitting the source and destination node indices (i.e., k and n), thereby reducing the variable to $y_{m,r,q}$. This abstraction shifts the focus from explicitly modeling end-to-end routes to merely ensuring that messages are transmitted through disjoint paths. Such a reduction in variable granularity can significantly streamline constraint formulation and lead to noticeable improvements in model generation time.

7 Conclusion and Outlook

This chapter concludes the thesis by summarizing the key contributions and findings of the research. It reflects on the strengths and limitations of the proposed work and outlines potential directions for future development. The discussion is structured to provide a comprehensive perspective on the work conducted, its practical implications, and how it can be extended to address remaining challenges in the field.

7.1 Conclusion

The design of automotive E/E architectures is facing unprecedented complexity due to the rapid growth in electronic systems, software-driven functionalities, and the increasing demands for safety, performance, and adaptability. Emerging technologies such as ADAS, vehicle connectivity, electrification, and smart mobility are continuously redefining the capabilities required of modern automotive systems. As the industry advances toward higher levels of vehicle autonomy, traditional hardware-centric development approaches are proving inadequate.

To address these limitations, the automotive domain is experiencing a paradigm shift toward zonal and centralized E/E architectures, which offer improved scalability, modularity, and integration. Simultaneously, vehicle software development has evolved to embrace service-oriented paradigms, enabling greater flexibility through hardware-agnostic software components. These components can be updated or extended throughout the vehicle's lifecycle, supporting the decoupled design of hardware and software elements [142].

A critical challenge in this context is the efficient and safe mapping of software components to hardware resources. Designing an architecture that satisfies stringent safety constraints—including timing requirements, fault tolerance,

D. Mohammad Zaheri, *Safety-Oriented Optimization of Automotive E/E Architectures*, Wissenschaftliche Reihe Fahrzeugtechnik Universität Stuttgart,
https://doi.org/10.1007/978-3-658-53045-7_7

and reliability—while also optimizing non-functional objectives such as latency and cost, is inherently complex. These objectives are often conflicting, and the trade-offs must be carefully managed to ensure a balanced and feasible solution.

In response to these challenges, this thesis introduces a comprehensive framework for automated design space exploration of automotive E/E architectures. The proposed approach facilitates the joint optimization of task mapping, ASIL decomposition, and message routing and scheduling, while adhering to ISO 26262 functional safety constraints. By automating key aspects of the design process, the framework significantly reduces manual engineering effort and improves both the quality and consistency of the final architecture.

This thesis addresses two key research questions in the design of mixed-criticality automotive systems.

The first question is: *"How can the allocation of mixed-criticality functions to hardware resources be optimized?"* To answer this, we proposed a MIP model that considers a multi-objective optimization targeting both development cost and end-to-end latency. Chapter 4 presents this model and highlights two distinct design strategies for automotive systems involving mixed ASILs. The first strategy, termed **Mixed-ASIL**, assumes that tasks (or subfunctions) with different ASIL levels can coexist on the same processing element without requiring an elevation of their individual ASIL ratings. This strategy relies on the execution environment to ensure the necessary freedom from interference (FFI) among co-located tasks. In contrast, the second strategy, referred to as **Max-ASIL**, assumes that such FFI guarantees are not available. Therefore, if tasks with different ASIL levels are assigned to the same processing element, the entire element must comply with the highest ASIL level among those tasks. This chapter formulates MIP-based solutions for both strategies, encompassing ASIL decomposition, function-to-resource mapping, and task scheduling. The decomposition approach adheres to the principles outlined in ISO 26262 and ensures compliance with key hardware architectural metrics. Experimental evaluations presented in the chapter confirm the effectiveness and applicability of the proposed optimization framework.

The second question explored in this thesis is: *"How should messages be routed and scheduled when ASIL decomposition is applied, and how can timing re-*

quirements be satisfied?" Chapter 5 introduces a model for optimizing message routing and scheduling in a mixed-criticality automotive network. The model considers the case where functions have been decomposed, and as a result, their corresponding messages may need to be sent multiple times—following the decomposition pattern. The solution respects FFI requirements as specified in ISO 26262, ensuring safe and interference-free communication.

The models and techniques presented in Chapters 4 and 5 can be integrated to form a unified framework for the design and optimization of mixed-criticality automotive E/E architectures. This integrated approach facilitates systematic design exploration, cost-efficient resource utilization, and compliance with functional safety standards.

7.2 Limitations and Challenges

While the presented framework demonstrates strong performance across multiple dimensions, several limitations remain that constrain its broader applicability and computational efficiency.

Computational Complexity

The proposed MIP formulation offers a highly expressive modeling approach, but this expressiveness comes at the cost of increased computational complexity. As the number of functions, ECUs, and communication requirements grows, the size of the problem increases rapidly. This is particularly evident when ASIL decomposition, task-to-resource allocation, task scheduling, and message routing are considered simultaneously, resulting in a highly constrained and large-scale optimization problem. Although strategies such as binary search heuristics and objective prioritization were employed to reduce solving time, these techniques only partially address the scalability challenge. For large automotive systems with tight timing and safety constraints, solving the complete model may become computationally intractable.

Static Assumptions

The current model assumes a static configuration at design time. Parameters such as failure rates, ECU availability, and workload profiles are fixed throughout the optimization. However, in real-world deployments, these parameters may change over time due to system aging, environmental variations, or runtime software updates. The absence of dynamic modeling reduces the framework's applicability in adaptive or reconfigurable system architectures, such as those supporting over-the-air updates or condition-based maintenance.

Limited Constraint Space

Although the framework addresses key objectives related to cost, safety, and latency, it does not account for several other critical system-level constraints. Factors such as power consumption, thermal limits, and energy-efficiency trade-offs have not been incorporated into the current model. These constraints are becoming increasingly significant in the context of electric and hybrid vehicles, where power-aware design is essential.

7.3 Future Work

To address the identified limitations and further extend the applicability of the proposed solution, future research will focus on two primary directions: (i) enhancing the optimization framework with richer system models and advanced heuristics to improve both scalability and realism, and (ii) integrating the framework into industrial toolchains to facilitate practical deployment and adoption.

- **Enhancing the Optimization Framework**

  One promising avenue for future development involves expanding the optimization framework to incorporate additional system-level objectives and constraints relevant to emerging trends in automotive system design.

Power and Energy Constraint: Incorporating detailed energy consumption models for both processing elements and communication links will enable the exploration of trade-offs between system performance and power efficiency. This extension is particularly critical for electric and hybrid vehicles, where energy-aware design decisions have a direct impact on vehicle range and thermal management.

Communication Time Heuristics: To address the scalability challenges associated with message routing and scheduling, heuristic-based methods—such as greedy algorithms, genetic algorithms, or reinforcement learning techniques—can be integrated either as pre-processing steps or as part of hybrid solving strategies. These approaches can significantly reduce computational overhead while maintaining near-optimal solution quality.

Task Reconfiguration Support: Future iterations of the framework should also support runtime reconfiguration and software updates. This includes modeling changes in workload, failure conditions, or system availability, thereby improving the robustness and adaptability of the solution in dynamic environments typical of software-defined vehicles.

Together, these enhancements will produce a more comprehensive and practical optimization tool, capable of addressing the increasing diversity of design requirements in modern E/E architectures.

- **Industrial Integration and Toolchain Support**

  Another key direction involves integrating the proposed framework with existing industrial tools and processes to support real-world automotive development.

  Toolchain Compatibility: Enabling model import and export compatibility with widely used platforms—such as PREEvision and AUTOSAR-compliant toolchains—will allow seamless integration of the optimization framework into existing engineering workflows. This compatibility will also facilitate the use of the proposed methods as a back-end optimization engine.

  User Interface and Visualization: Developing a graphical user interface (GUI) will improve accessibility and usability, enabling users to visualize and interact with the function dependency graph, resource mappings, and system-level metrics in an intuitive manner.

Co-simulation and Validation: Integrating the framework with existing timing and safety co-simulation tools will support end-to-end validation and verification, thereby closing the loop between architectural optimization and compliance assessment.

Open-source Deployment: Finally, releasing the framework as an open-source project can encourage community contributions, support broader evaluation, and enable continuous improvement based on industry and academic feedback.

These integration efforts will significantly enhance the framework's accessibility and practical utility, facilitating its use in both early-stage design exploration and final system validation phases in automotive development lifecycles.

Glossary

ADAS	Advanced Driver Assistance Systems, such as lane keeping assist, emergency braking, or adaptive cruise control
DO-178C	Software Considerations in Airborne Systems and Equipment Certification, a guideline published by RTCA for ensuring safety and reliability in the development of aviation software, widely used in the aerospace industry
EN 50128	European standard for software development in railway control and protection systems, specifying requirements for ensuring functional safety in programmable electronic systems
IEC 61508	International standard for the functional safety of electrical, electronic, and programmable electronic safety-related systems, published by the International Electrotechnical Commission (IEC), first issued in 1998 and revised in 2010
IEC 62061	International standard for the functional safety of electrical, electronic, and programmable electronic control systems in machinery, based on IEC 61508 and focused on machinery-specific safety requirements
IEC 62304	International standard specifying life cycle requirements for the development of medical device software, aligned with IEC 61508 and focused on software safety classification and risk management
ISO 26262	International standard for functional safety of electrical and electronic systems in the automotive domain, defined by the International Organization for Standardization (ISO) in 2011 and revised in 2018

D. Mohammad Zaheri, *Safety-Oriented Optimization of Automotive E/E Architectures*, Wissenschaftliche Reihe Fahrzeugtechnik Universität Stuttgart,
https://doi.org/10.1007/978-3-658-53045-7

SOTIF — Safety of the Intended Functionality, a standard (ISO/PAS 21448) addressing hazards resulting from functional insufficiencies or performance limitations of the intended functionality, especially relevant for advanced driver assistance systems (ADAS) and autonomous driving

UNICARagil — A German research project initiated in 2018, aiming to develop a disruptive, modular vehicle platform for fully autonomous driving, integrating centralized architecture and innovative driving dynamics concepts

Bibliography

[1] Shugang Jiang. Vehicle E/E Architecture and Its Adaptation to New Technical Trends. *SAE Tech. Paper*, 2019. doi: 10.4271/2019-01-0862.

[2] Klaus Grimm. Software technology in an automotive company - major challenges. *25th International Conference on Software Engineering*, pages 498–503, 2003. doi: 10.1109/ICSE.2003.1201228.

[3] Laura Sophie Brandt. *Architekturgesteuerte Elektrik/Elektronik Baukastenentwicklung im Automobil*. PhD Thesis, Fakultät für Maschinenwesen, Technische Universität München, 2016.

[4] Wenwei Wang, Kaidi Guo, Wanke Cao, Hailong Zhu, Jinrui Nan, and Lei Yu. Review of Electrical and Electronic Architectures for Autonomous Vehicles: Topologies, Networking and Simulators. *Automotive Innovation*, 7(1):82–101, 2024. ISSN 2096-4250. doi: 10.1007/s42154-023-0 0266-9.

[5] Johannes David Eder. *Automatic Exploration of Automotive E/E Architectures*. PhD Thesis, Technischen Universität München, 2022. URL `https://mediatum.ub.tum.de/?id=1638880`.

[6] Sangeeth Kochanthara, Niels Rood, Arash Khabbaz Saberi, Loek Cleophas, Yanja Dajsuren, and Mark van den Brand. A functional safety assessment method for cooperative automotive architecture. *Journal of Systems and Software*, 179:110991, 2021. ISSN 01641212. doi: 10.1016/j.jss.2021.110991.

[7] Lucas Mauser and Stefan Wagner. Centralization potential of automotive E/E architectures. *Journal of Systems and Software*, 219:112220, 2025. ISSN 01641212. doi: 10.1016/j.jss.2024.112220.

D. Mohammad Zaheri, *Safety-Oriented Optimization of Automotive E/E Architectures*, Wissenschaftliche Reihe Fahrzeugtechnik Universität Stuttgart, https://doi.org/10.1007/978-3-658-53045-7

[8] Varun M. Navale, Kyle Williams, Athanassios Lagospiris, Michael Schaffert, and Markus-Alexander Schweiker. (R)evolution of E/E Architectures. *SAE International Journal of Passenger Cars - Electronic and Electrical Systems*, 8(2):282–288, 2015. ISSN 1946-4622. doi: 10.4271/2015-01-0196.

[9] Hadi Askaripoor, Morteza Hashemi Farzaneh, and Alois Knoll. E/E Architecture Synthesis: Challenges and Technologies. *Electronics*, 11 (4):518, 2022. doi: 10.3390/electronics11040518.

[10] FKFS- Research Institute for Automotive Engineering and Powertrain Systems Stuttgart. E/E Architecture Optimization. URL `https://www.fkfs.de/en/news/news/detail/highlights-at-fkfs-e-e-architecture-optimization`.

[11] Victor Bandur, Gehan Selim, Vera Pantelic, and Mark Lawford. Making the Case for Centralized Automotive E/E Architectures. *IEEE Transactions on Vehicular Technology*, 70(2):1230–1245, 2021. ISSN 0018-9545. doi: 10.1109/TVT.2021.3054934.

[12] Jochen Klaus-Wagenbrenner. Zonal EE Architecture: Towards a Fully Automotive Ethernet–Based Vehicle Infrastructure. 2019. URL `https://standards.ieee.org/wp-content/uploads/import/documents/other/eipatd-presentations/2019/D1-04_KLAUS-Zonal_EE_Architecture.pdf`.

[13] Thomas Liebetrau. E/E Architecture Transformation How it impacts value chain and networking technologies. *AmE 2022 - Automotive meets Electronics; 13. GMM-Symposium, Dortmund, Germany*, 104, 2022.

[14] Hyuksoo Jang, Chulsun Park, Sangho Goh, and Sungkwon Park. Design of a Hybrid In-Vehicle Network Architecture Combining Zonal and Domain Architectures for Future Vehicles. *IEEE 6th International Conference on Knowledge Innovation and Invention (ICKII)*, pages 33–37, 2023. doi: 10.1109/ICKII58656.2023.10332574.

[15] McKinsey & Company and GSA. Getting ready for next-generation E/E architecture with zonal compute, 2023. URL https://www.mckinsey.com/industries/semiconductors/our-insights/getting-ready-for-next-generation-ee-architecture-with-zonal-compute.

[16] Azianti Ismail and Won Jung. International Trend of Functional Safety. *Korea Society of Industrial and Systems Engineering*, 2013.

[17] IEC 61508. Functional safety of electrical/electronic/programmable electronic safety-related systems, 2010.

[18] ISO 26262. International Organization for Standardization. Road vehicles - functional safety, 2018.

[19] ISO 21448. International Organization for Standardization. Road vehicles — Safety of the intended functionality, 2022.

[20] DO 178C/ED-12C. European Organisation for Civil Aviation Equipment. Software Considerations in Airborne Systems and Equipment Certification., 2012.

[21] EN 50128. European Committee for Electrotechnical Standardization. Railway applications - communication, signalling and processing systems - Software for railway control and protection systems, 2020.

[22] IEC 62304. International Electrotechnical Commission. Medical device software - Software life cycle processes, 2006.

[23] IEC 62061. International Electrotechnical Commission . Safety of machinery - Functional safety of safety-related control systems, 2021.

[24] Timo Woopen, Bastin Lampe, and et al. UNICARagil - Disruptive Modular Architectures for Agile, Automated Vehicle Concepts. *27th Aachen Colloquium Automobile and Engine Technology*, 2018.

[25] Mario Hirz. An approach supporting integrated modeling and design of complex mechatronics products by the example of automotive applications. *22nd Multi-Conference on Systemics, Cybernetics and Informatics - WMSCI*, 2018.

[26] Dorsa Zaheri and Hans-Christian Reuss. Considering Safety Requirements Based on ISO 26262 in Model-Based Optimization of Automotive Electrical/Electronic Architectures. *Journal Of Tongji University (Natural Science)*, 2022. doi: 10.11908/j.issn.0253-374x.22-SSS1-23751.

[27] Rakesh Rana, Miroslaw Staron, Christian Berger, Jörgen Hansson, Martin Nilsson, and Fredrik Törner. Early Verification and Validation According to ISO 26262 by Combining Fault Injection and Mutation Testing. *In: Cordeiro, J., van Sinderen, M. (eds) Software Technologies. ICSOFT 2013. Communications in Computer and Information Science*, 457:164–179, 2014. doi: 10.1007/978-3-662-44920-2{\textunderscore}11.

[28] Peter Waszecki, Martin Lukasiewycz, Alejandro Masrur, and Samarjit Chakraborty. How to engineer tool-chains for automotive E/E architectures? *ACM Sigbed Review*, 10(4):6–15, 2013. doi: 10.1145/2583687.2583689.

[29] SDV101-Part D: Implementation Strategies. The Traditional V-Model in Automotive Development. URL `https://www.sdv.guide/sdv101/part-d-implementation-strategies/hardware-vs-software-engineering/the-traditional-v-model-in-automotive-development`.

[30] Stefan Kugele and Gheorghe Pucea. Model-based optimization of automotive E/E-architectures. *Proceedings of the 6th International Workshop on Constraints in Software Testing, Verification, and Analysis (CSTVA)*, pages 18–29, 2014. doi: 10.1145/2593735.2593739.

[31] Hassan Hage, Vahid Hashemi, and Frank Mantwill. Towards a Systems Engineering based Automotive Product Engineering Process. *Automotive System/Software Architectures Workshop @ECSA*, 2020. doi: 10.48550/ARXIV.2007.11897.

[32] Vector. PREEvision, 2021. URL `https://www.vector.com/de/en/products/products-a-z/software/preevision/#c1789`.

[33] Jörg Schäuffele. E/E Architectural Design and Optimization using PREEvision. *SAE Technical Paper*, 2016. doi: 10.4271/2016-01-0016.

[34] Vector Informatik GmbH. PREEvision- User Manual: Version 10.0, 2022.

[35] The Community for Open Collaboration and Innovation:. The Eclipse Foundation. URL https://www.eclipse.org/.

[36] Simon Fürst, J. Mössinger, S. Bunzel, T. Weber, F. Kirschke-Biller, P. Heitkämper, G. Kinkelin, K. Nishikawa, and K. Lange. AUTOSAR – A Worldwide Standard is on the Road. *In Proceedings of the 14th International VDI Congress Electronic Systems for Vehicles, Baden-Baden*, 2009.

[37] Volcano Vehicle Systems Architect. URL https://de.mathworks.com/products/connections/product_detail/volcano-vehicle-systems-architect.html.

[38] Unified Modeling Language. URL https://www.uml.org/.

[39] Systems Modeling Language. URL https://sysml.org/.

[40] IBM Engineering Rhapsody Designer for Systems Engineers. URL https://www.ibm.com/products/engineering-rhapsody-designer.

[41] Vector. Analyzing Automotive ECUs and Networks With CANalyzer. URL https://www.vector.com/se/en/products/products-a-z/software/canalyzer/#c3451.

[42] Elektrobit. Classic AUTOSAR with EB tresos. URL https://www.elektrobit.com/products/ecu/eb-tresos/.

[43] MathWorks. EB tresos Studio and EB tresos AutoCore. URL https://de.mathworks.com/products/connections/product_detail/eb-tresos.html.

[44] ASCET-DEVELOPER. ETAS. URL https://www.etas.com/en/products/ascet-developer.php.

[45] AUTOSAR Builder. Welcome to AUTOSAR Builder 2020x. URL `https://www.3ds.com/fileadmin/Welcome_to_AUTOSAR_Builder_2020x.pdf`.

[46] Autosarbuilder—Dassault Systèmes. URL `https://www.3ds.com/products/catia/autosar-builder`.

[47] chronSIM INCHRON:. Unleash the Power of Timing Simulation. URL `https://www.inchron.com/chronsim/`.

[48] chronVAL INCHRON:. Worst-Case Timing Analysis. URL `https://www.inchron.com/chronval/`.

[49] Saoussen Anssi, Karsten Albers, Matthias Dörfel, and Sébastien Gérard. ChronVAL/ChronSIM: A Tool Suite for Timing Verification of Automotive Applications. *Embedded Real Time Software and Systems (ERTS)*, 2012. URL `https://hal.science/hal-02191852/document`.

[50] R. Henia, A. Hamann, M. Jersak, R. Racu, K. Richter, and R. Ernst. System level performance analysis – the SymTA/S approach. *IEE Proceedings - Computers and Digital Techniques*, 152(2):148, 2005. ISSN 13502387. doi: 10.1049/ip-cdt:20045088.

[51] Arne Hamann, Rafik Henia, Razvan Racu, Marek Jersak, Kai Richter, and Rolf Ernst. SymTA/S - Symbolic Timing Analysis for Systems. *In Proceedings of the WIP Proc. Euromicro Conference on Real-Time Systems 2004 (ECRTS'04)*, 2004.

[52] Elizabeth Hull, Ken Jackson, and Jeremy Dick. DOORS: A Tool to Manage Requirements. *Requirements Engineering. Practitioner Series. Springer, London*, 2002. doi: 10.1007/978-1-4471-3730-6{\textunderscore}9.

[53] Arash Khabbaz Saberi. *Functional Safety: A New Architectural Perspective: Model-based safety engineering for automated driving systems.* PhD Thesis, Eindhoven University of Technology, 2020.

[54] Azianti Ismail and Won Jung. Research trends in automotive functional safety. *International Conference on Quality, Reliability, Risk, Maintenance, and Safety Engineering (QR2MSE)*, pages 1–4, 2013. doi: 10.1109/QR2MSE.2013.6625523.

[55] Yaping Luo. *From conceptual models to safety assurance : applying model-based techniques to support safety assurance*. PhD Thesis, Mathematics and Computer Science, Technische Universiteit Eindhoven, 2016.

[56] D.D Ward and S.E Crozier. The uses and abuses of ASIL decomposition in ISO 26262. *7th IET International Conference on System Safety, incorporating the Cyber Security Conference*, 2012. doi: 10.1049/cp.2012.1523.

[57] Software for Safety-Related Automotive Systems: Best Practice Guideline. *ZVEI-German Electrical and Electronic Manufacturers' Association*, 2016. URL https://www.zvei.org/fileadmin/user_upload/Presse_und_Medien/Publikationen/2016/August/Software_for_Safety-Related_Automotive_Systems__Best_Practice_Guideline_/Software-for-Safety-Related-Automotive-Systems-Guideline-ZVEI-Stand-2016-12.pdf.

[58] Nabarun Das and William Taylor. Quantified fault tree techniques for calculating hardware fault metrics according to ISO 26262. *IEEE Symposium on Product Compliance Engineering (ISPCE)*, pages 1–8, 2016. doi: 10.1109/ISPCE.2016.7492848.

[59] Kuen-Long Lu and Yung-Yuan Chen. Safety-Oriented System Hardware Architecture Exploration in Compliance with ISO 26262. *Applied Sciences*, 12(11):5456, 2022. doi: 10.3390/app12115456.

[60] Abraham Cherfi. *Toward an Efficient Generation of ISO 26262 Automotive Safety Analyses*. PhD Thesis, Ecole Doctorale Polytechnique, Computer Science [cs], 2015.

[61] Arne Hamann. *Iterative design space exploration and robustness optimization for embedded systems*. PhD Thesis, TU Braunschweig, Göttingen, 2008.

[62] Stefan Kugele, Philipp Obergfell, and Eric Sax. Model-based resource analysis and synthesis of service-oriented automotive software architectures. *Software and Systems Modeling*, 20(6):1945–1975, 2021. doi: 10.1007/s10270-021-00896-9.

[63] Amit Kumar Singh, Muhammad Shafique, Akash Kumar, and Jörg Henkel. Mapping on multi/many-core systems: Survey of current and emerging trends. *50th ACM/EDAC/IEEE Design Automation Conference (DAC)*, pages 1–10, 2013. doi: 10.1145/2463209.2488734.

[64] Manjari Gupta, Lava Bhargava, and S. Indu. Mapping techniques in multicore processors: current and future trends. *The Journal of Supercomputing*, 77(8):9308–9363, 2021. doi: 10.1007/s11227-021-03650-6.

[65] Aldeida Aleti, Stefan Bjornander, Lars Grunske, and Indika Meedeniya. ArcheOpterix: An extendable tool for architecture optimization of AADL models. *ICSE Workshop on Model-Based Methodologies for Pervasive and Embedded Software*, pages 61–71, 2009. doi: 10.1109/MOMPES.2009.5069138.

[66] Issam Al-Azzoni, Julian Blank, and Nenad Petrović. A Model-Driven Approach for Solving the Software Component Allocation Problem. *Algorithms*, 14(12):354, 2021. doi: 10.3390/a14120354.

[67] Indika Meedeniya, Barbora Buhnova, Aldeida Aleti, and Lars Grunske. Reliability-driven deployment optimization for embedded systems. *Journal of Systems and Software*, 84(5):835–846, 2011. ISSN 01641212. doi: 10.1016/j.jss.2011.01.004.

[68] Steffen Becker, Heiko Koziolek, and Ralf Reussner. The Palladio component model for model-driven performance prediction. *Journal of Systems and Software*, 82(1):3–22, 2009. ISSN 01641212. doi: 10.1016/j.jss.2008.03.066.

[69] Anne Koziolek, Heiko Koziolek, and Ralf Reussner. PerOpteryx: Automated Application of Tactics in Multi-Objective Software Architecture Optimization. *7th International Conference on the Quality of Software Architectures, QoSA 2011 and 2nd International Symposium on Architecting Critical Systems, ISARCS*, 2011. doi: 10.1145/2000259.

[70] Vincent Aravantinos, Sebastian Voss, Sabine Teufl, Florian Hölzl, and Bernhard Schätz. AutoFOCUS 3: Tooling Concepts for Seamless, Model-based Development of Embedded Systems. *8th International Workshop on Model-based Architecting of Cyber-Physical and Embedded Systems*, 2015.

[71] Sebastian Voss and Bernhard Schätz. Scheduling shared memory multicore architectures in A UTO FOCUS 3 using Satisfiability Modulo Theories. *In Modellbasierte Entwicklung eingebetteter Systeme*, 2012.

[72] Sebastian Voss and Bernhard Schätz. Deployment and Scheduling Synthesis for Mixed-Critical Shared-Memory Applications. *20th Annual IEEE International Conference and Workshops on the Engineering of Computer Based Systems (ECBS)*, pages 100–109, 2013. doi: 10.1109/ECBS.2013.23.

[73] Sebastian Voss, Johannes Eder, and Florian Hölzl. Design Space Exploration and its Visualization in AUTOFOCUS3. *Software Engineering*, 2014.

[74] Thomas L. Saaty. *Fundamentals of Decision Making and Priority Theory With the Analytic Hierarchy Process*. Rws Publications, 1994.

[75] Ivan Švogor and Jan Carlson. SCALL: Software Component Allocator for Heterogeneous Embedded Systems. *In Proceedings of the 2015 European Conference on Software ArchitectureWorkshops*, pages 1–5, 2015. doi: 10.1145/2797433.2797501.

[76] Hadi Askaripoor, Thilo Mueller, and Alois Knoll. E/E Designer: A Framework to Design and Synthesize Vehicle E/E Architecture. *IEEE Transactions on Intelligent Vehicles*, 9(7):5204–5221, 2024. doi: 10.110 9/TIV.2023.3324617.

[77] Athanasios Retouniotis, Yiannis Papadopoulos, and Ioannis Sorokos. Andromeda: A model-connected framework for safety assessment and assurance. *Journal of Systems and Software*, 220:112256, 2025. ISSN 01641212. doi: 10.1016/j.jss.2024.112256.

[78] Ken W. Tindell, Alan Burns, and Andy J. Wellings. Allocating hard real-time tasks: An NP-Hard problem made easy. *Journal of Real-Time Systems*, 1992.

[79] Klaus Becker. *Software Deployment Analysis for Mixed Reliability Automotive Systems*. PhD Thesis, Technische Universität München, 2017. URL `https://mediatum.ub.tum.de/1345914`.

[80] Fengjunjie Pan, Markus Rickert, Tobias Betz, Long Wen, Jianjie Lin, Nenad Petrovic, Markus Lienkamp, and Alois Knoll. Towards Software-Defined Vehicles: From Model-based Engineering to Virtualization-based Deployment. *IEEE Access*, page 1, 2024. doi: 10.1109/ACCESS .2024.3512002.

[81] Fengjunjie Pan, Jianjie Lin, Markus Rickert, and Alois Knoll. Automated Design Space Exploration for Resource Allocation in Software-Defined Vehicles. *IEEE Intelligent Vehicles Symposium (IV)*, pages 1–8, 2023. doi: 10.1109/IV55152.2023.10186605.

[82] L. de Moura and N. Bjørner. Z3: An Efficient SMT Solver. *In Tools and Algorithms for the Construction and Analysis of Systems (TACAS)*, 2008.

[83] Fengjunjie Pan, Jianjie Lin, Markus Rickert, and Alois Knoll. Resource Allocation in Software-Defined Vehicles: ILP Model Formulation and Solver Evaluation. *IEEE 25th International Conference on Intelligent Transportation Systems (ITSC)*, 2022.

[84] Alexandru Kampmann, Maximilian Lüer, Stefan Kowalewski, and Bassam Alrifaee. Optimization-based Resource Allocation for an Automotive Service-oriented Software Architecture. *IEEE Intelligent Vehicles Symposium (IV)*, 2022.

[85] Alexandru Kampmann, Bassam Alrifaee, Markus Kohout, Andreas Wüstenberg, Timo Woopen, Marcus Nolte, Lutz Eckstein, and Stefan Kowalewski. A Dynamic Service-Oriented Software Architecture for Highly Automated Vehicles. *IEEE Intelligent Transportation Systems Conference (ITSC)*, 2019.

[86] Pierre-Emmanuel Hladik, Hadrien Cambazard, Anne-Marie Déplanche, and Narendra Jussien. Solving a real-time allocation problem with constraint programming. *Journal of Systems and Software*, 81(1):132–149, 2008. ISSN 01641212. doi: 10.1016/j.jss.2007.02.032.

[87] Bas Boone, Filip de Turck, and Bart Dhoedt. Automated Deployment of Distributed Software Components with Fault Tolerance Guarantees. *Sixth International Conference on Software Engineering Research, Management and Applications*, pages 21–27, 2008. doi: 10.1109/SERA.2008.28.

[88] Qi Zhu, Haibo Zeng, Wei Zheng, Marco Di Natale, and Alberto Sangiovanni-Vincentelli. Optimization of task allocation and priority assignment in hard real-time distributed systems. *ACM Transactions on Embedded Computing Systems*, 11(4):1–30, 2012. doi: 10.1145/2362336.2362352.

[89] Bernd Hardung. *Optimisation of the Allocations of Functions in Vehicle Networks*. PhD Thesis, University of Erlangen-Nuremberg, 2006.

[90] Martin Lukasiewycz, Michael Glass, Christian Haubelt, Jurgen Teich, Richard Regler, and Bardo Lang. Concurrent Topology and Routing Optimization in Automotive Network Integration. *45th ACM/IEEE Design Automation Conference*, 2008. doi: 10.1145/1391469.1391629.

[91] Martin Lukasiewycz. *Modeling, analysis, and optimization of automotive networks*. PhD Thesis, Ingolstadt Institute of Friedrich-Alexander-University Erlangen-Nuremberg and AUDI AG (INI.FAU), 2010.

[92] Raphael Weber, Stefan Henkler, and Achim Rettberg. Multi–objective design space exploration for cyber–physical systems satisfying hard real–time and reliability constraints. *In Proceedings of IDEAL'14 Workshop, IFIP Springer Series*, 2014.

[93] Florian Pölzlbauer, Iain Bate, and Eugen Brenner. Software Deployment for Distributed Software Deployment for Distributed Embedded Real-Time Systems of Automotive Applications. *In Embedded and Real Time System Development: A Software Engineering Perspective*, 2014.

[94] Johannes Eder, Sebastian Voss, Andreas Bayha, Alexandru Ipatiov, and Maged Khalil. Hardware architecture exploration: automatic exploration of distributed automotive hardware architectures. *Software and Systems Modeling*, 19(4):911–934, 2020. doi: 10.1007/s10270-020-00786-6.

[95] Tarik Terzimehic´, Simon Barner, Yuri Gil Dantas, Ulrich Schopp, Vivek Nigam, and Pei Ke. Safety-Aware Deployment Synthesis and Trade-Off Analysis of Apollo Autonomous Driving Platform. *IEEE 20th International Conference on Software Architecture Companion (ICSA-C)*, 2023.

[96] Yuri Dantas, Tiziano Munaro, Carmen Carlan, Vivek Nigam, Simon Barner, Shiqing Fan, Alexander Pretschner, Ulrich Schöpp, and Sergey Tverdyshev. A Model-based System Engineering Plugin for Safety Architecture Pattern Synthesis. *In 10th International Conference on Model-Driven Engineering and Software Development*, pages 36–47, 2022. doi: 10.5220/0010831700003119.

[97] Sally C. Brailsford, Chris N. Potts, and Barbara M. Smith. Constraint satisfaction problems: Algorithms and applications. *European Journal of Operational Research*, 1999. doi: 10.1016/S0377-2217(98)00364-6.

[98] Joaquim R. R. A. Martins and S. Andrew Ning. *Engineering design optimization*. Cambridge University Press, Cambridge and New York NY, 2021.

[99] Hermann Kopetz. *Real-Time Systems: Design Principles for Distributed Embedded Applications*. Springer, 2011.

[100] Dorsa Zaheri, Dennis Niedballa, Marc Leuffen, and Gergely Bilkei-Gorzo. Practical Implementation of a Zonal E/E Architecture for Autonomous Vehicles in UNICARagil. *ATZ Electron Worldw*, 2023. doi: 10.1007/s38314-023-1436-x.

[101] Mohamed Slim Dhouibi, Jean-Marc Perquis, Laurent Saintis, and Mihaela Barreau. Automatic Decomposition and Allocation of Safety Integrity Level Using System of Linear Equations. *in Proceedings of the Fourth International Conference on Performance, Safety and Robustness in Complex Systems and Applications*, 2014. doi: 10.13140/2.1.2856.0321.

[102] Jaroslav Menčík. *Concise Reliability for Engineers*. InTech, 2016. doi: 10.5772/62009.

[103] Youcef Gheraibia, Sohag Kabir, Khaoula Djafri, and Habiba Krimou. An Overview of the Approaches for Automotive Safety Integrity Levels Allocation. *Journal of Failure Analysis and Prevention*, 18(3):707–720, 2018. doi: 10.1007/s11668-018-0466-9.

[104] System Composer. Model-Based Systems Engineering. URL `https://de.mathworks.com/solutions/model-based-systems-engineering.html`.

[105] Domitian Tamas-Selicean. *Design of Mixed-Criticality Applications on Distributed Real-Time Systems*. PhD Thesis, Technical University of Denmark, 2015.

[106] Paul Pop Domit¸ian Tamas¸–Selicean. Design Optimization of Mixed-Criticality Real-Time Embedded Systems. *ACM Transactions on Embedded Computing Systems (TECS)*, 2015. doi: 10.1145/2700103.

[107] Junhe Gan. *Tradeoff Analysis for Dependable Real-Time Embedded Systems during the Early Design Phases*. PhD. dissertation, Department of Applied Mathematics and Computer Science, Technical University of Denmark, 2014.

[108] Guoqi Xie, Yuekun Chen, Yan Liu, Renfa Li, and Keqin Li. Minimizing Development Cost With Reliability Goal for Automotive Functional Safety During Design Phase. *IEEE Transactions on Reliability*, 67(1): 196–211, 2018. doi: 10.1109/TR.2017.2778070.

[109] Guoqi Xie, Wei Wu, Gang Zeng, Renfa Li, and Shiyan Hu. Risk Assessment and Development Cost Optimization in Software Defined Vehicles. *IEEE Transactions on Intelligent Transportation Systems*, 22(6):3675–3686, 2021. doi: 10.1109/TITS.2020.3027469.

[110] Guoqi Xie, Hao Peng, Jing Huang, Renfa Li, and Keqin Li. Energy-Efficient Functional Safety Design Methodology Using ASIL Decomposition for Automotive Cyber-Physical Systems. *IEEE Transactions on Reliability*, pages 1–23, 2019. doi: 10.1109/TR.2019.2915818.

[111] Biao Hu, Shengjie Xu, Zhengcai Cao, and Mengchu Zhou. Safety-Guaranteed and Development Cost- Minimized Scheduling of DAG Functionality in an Automotive System. *IEEE Transactions on Intelligent Transportation Systems*, 23(4):3074–3086, 2022. doi: 10.1109/TITS.2020.3030607.

[112] Alessandro Frigerio, Bart Vermeulen, and Kees Goossens. A Generic Method for a Bottom-Up ASIL Decomposition. *In: Gallina, B., Skavhaug, A., Bitsch, F. (eds) Computer Safety, Reliability, and Security. SAFECOMP. Lecture Notes in Computer Science*, 2018. doi: 10.1007/978-3-319-99130-6{\textunderscore}2.

[113] Alessandro Frigerio, Bart Vermeulen, and Kees Goossens. Component-Level ASIL Decomposition for Automotive Architectures. *in Proc. 49th Annu. IEEE/IFIP Int. Conf. Dependable Syst. Netw. Workshops (DSN-W)*, pages 62–69, 2019. doi: 10.1109/DSN-W.2019.00021.

[114] Alessandro Frigerio. *Functional-safety analysis of ASIL decomposition for redundant automotive systems*. PhD Thesis, Eindhoven University of Technology, 2022.

[115] Mohamed Slim Dhouibi. *Optimization approach for the critical automotive embedded systems*. PhD Thesis, Université d'Angers, 2016.

[116] Dorsa Zaheri and Hans-Christian Reuss. ASIL-Decomposition Based Resource Allocation Optimization for Automotive E/E Architectures. *IEEE Intelligent Vehicles Symposium (IV)*, 2025.

[117] AUTOSAR. Overview of Functional Safety Measures in AUTOSAR, 2016. URL `https://www.autosar.org/fileadmin/standards/R4.3.1/CP/AUTOSAR_EXP_FunctionalSafetyMeasures.pdf`.

[118] Sofiene Beji, Sardaouna Hamadou, Abdelouahed Gherbi, and John Mullins. SMT-Based Cost Optimization Approach for the Integration of Avionic Functions in IMA and TTEthernet Architectures. *IEEE/ACM 18th International Symposium on Distributed Simulation and Real Time Applications*, pages 165–174, 2014. doi: 10.1109/DS-RT.2014.28.

[119] Hossein Arsham. Big-M free solution algorithm for general linear programs. *In: International Journal of Pure and Applied Mathematics 32.4*, 2006.

[120] Ikram Senoussaoui, Houssam-Eddine Zahaf, Giuseppe Lipari, and Kamel Benhaoua. Contention-free scheduling of PREM tasks on partitioned multicore platforms. *IEEE 27th International Conference on Emerging Technologies and Factory Automation (ETFA)*, 2022.

[121] Mohammad Asghari, Amir M. Fathollahi-Fard, S. M. J. Mirzapour Al-e-hashem, and Maxim A. Dulebenets. Transformation and Linearization Techniques in Optimization: A State-of-the-Art Survey. *Mathematics*, 10(2):283, 2022. doi: 10.3390/math10020283.

[122] Gurobi Optimization L. L.C. Gurobi Optimizer Reference Manual, 2024. URL `https://www.gurobi.com`.

[123] David Parker, Martin Walker, Luis Silva Azevedo, Yiannis Papadopoulos, and Rui Esteves Araújo. Automatic Decomposition and Allocation of Safety Integrity Levels Using a Penalty-Based Genetic Algorithm. *In Recent Trends in Applied Artificial Intelligence*, 2013. doi: 10.1007/978-3-642-38577-3{\textunderscore}46.

[124] Roland Mader, Eric Armengaud, Andrea Leitner, and Christian Steger. Automatic and optimal allocation of safety integrity levels. *Proceedings Annual Reliability and Maintainability Symposium, Reno, NV, USA*, pages 1–6, 2012. doi: 10.1109/RAMS.2012.6175431.

[125] Luis Silva Azevedo, David Parker, Martin Walker, Yiannis Papadopoulos, and Rui Esteves Araùjo. Automatic Decomposition of Safety Integrity Levels: Optimization by Tabu Search. *SAFECOMP- Workshop CARS (2nd Workshop on Critical Automotive applications : Robustness & Safety) of the 32nd International Conference on Computer Safety, Reliability and Security*, 2013.

[126] Subba Rao V. Majety, Milind Dawande, and Jayant Rajgopal. Optimal Reliability Allocation with Discrete Cost-Reliability Data for Components. *In Operations Research*, 1999.

[127] Alain Billionnet. Redundancy Allocation for Series-Parallel Systems Using Integer Linear Programming. *IEEE Transactions on Reliability*, 57(3):507–516, 2008. doi: 10.1109/TR.2008.927807.

[128] Patrizio Pelliccione, Eric Knauss, Rogardt Heldal, S. Magnus Ågren, Piergiuseppe Mallozzi, Anders Alminger, and Daniel Borgentun. Automotive Architecture Framework: The experience of Volvo Cars. *Journal of Systems Architecture*, 77:83–100, 2017. ISSN 13837621. doi: 10.1016/j.sysarc.2017.02.005.

[129] Jonas Maier and Hans-Christian Reuss. Handling System Complexity in Zonal E/E Architectures. *Transportation Engineering*, 13:100195, 2023. ISSN 2666691X. doi: 10.1016/j.treng.2023.100195.

[130] Seongwoo Kim, Eundong Lee, Mideum Choi, Hanyou Jeong, and Seungwoo Seo. Design Optimization of Vehicle Control Networks. *IEEE Transactions on Vehicular Technology*, 60(7):3002–3016, 2011. doi: 10.1109/TVT.2011.2161680.

[131] Jörg Sommer and Elias A. Doumith. Topology Optimization of In-vehicle Multimedia Communication Systems. *1st International ICST Symposium on Vehicular Computing Systems*, 2008.

[132] Taeju Park. *Optimization of In-Vehicle Network Design*. PhD Thesis, Computer Science and Engineering, 2021.

[133] Thilo Streichert, Christian Haubelt, and Jurgen Teich. Multi-Objective Topology Optimization for Networked Embedded Systems. *International Conference on Embedded Computer Systems*, pages 93–98, 2006. doi: 10.1109/ICSAMOS.2006.300814.

[134] Ralph Moritz, Tamara Ulrich, and Lothar Thiele. Evolutionary Exploration of E/E-Architectures in Automotive Design. *Operations Research Proceedings*, 2012. doi: 10.1007/978-3-642-29210-1{\textunderscore}5 8.

[135] Martin Lukasiewycz, Michael Glass, Christian Haubelt, and Jurgen Teich. Efficient symbolic multi-objective design space exploration. *In Proceedings of the 13th Asia and South Pacific Design Automation*, pages 691–696, 2008. doi: 10.1109/ASPDAC.2008.4484040.

[136] Hadi Askaripoor, Morteza Hashemi Farzaneh, and Alois Knoll. A Model-Based Approach to Facilitate Design of Homogeneous Redundant E/E Architectures. *IEEE Intelligent Transportation Systems Conference (ITSC)*, pages 3426–3431, 2021. doi: 10.1109/ITSC48978.2021.95651 15.

[137] Lucas Haug. *Optimizing ILP-based Joint Scheduling and Routing for Time-Aware Shaping in Factory Automation Networks*. Bachelor thesis, University of Stuttgart, Institute of Parallel and Distributed Systems, 2020.

[138] Martin Dohr. *Automotive Network Optimization*. PhD Thesis, Graz University of Technology, 2014.

[139] Yuanbin Zhou, Soheil Samii, Petru Eles, and Zebo Peng. ASIL-Decomposition Based Routing and Scheduling in Safety-Critical Time-Sensitive Networking. *IEEE 27th Real-Time and Embedded Technology and Applications Symposium (RTAS)*, pages 184–195, 2021. doi: 10.1109/RTAS52030.2021.00023.

[140] Hadi Askaripoor. *A Framework to Design and Synthesize Vehicle E/E Architecture*. PhD Thesis, Technische Universität München, 2024.

[141] P. Erdös and A. Rényi. On random graphs I. *Publ. Math. Debrecen*, 1959.

[142] Houssem Guissouma, Heiko Klare, Eric Sax, and Erik Burger. An Empirical Study on the Current and Future Challenges of Automotive Software Release and Configuration Management. *In 44th Euromicro Conference on Software Engineering and Advanced Applications*, pages 298–305, 2018. doi: 10.1109/SEAA.2018.00056.

Appendix

A. Appendix A

Code A.1: Generation of Initial Population and ECU Assignment

```
# Generate initial population where each task randomly selects a decomposition scheme
      and assigns ECUs
def generate_initial_population():
    population = []
    for _ in range(population_size):
        individual = []
        for _ in range(len(tasks)):
            scheme = random.choice(decomposition_schemes)
            assignments = assign_subtasks_to_ecus(scheme)
            individual.append((scheme, assignments)) # Tuple: (scheme, [(subtask, ECU)])
        population.append(individual)
    return population
# Function to assign specific ECUs to each subtask within the decomposition scheme
def assign_subtasks_to_ecus(scheme):
    subtasks = []
    available_ecus = hardware.copy()
    # Breaking down each scheme's subtask assignment for better line control
    scheme_mappings = {
        '4xA': [('A',
              available_ecus.pop(available_ecus.index(random.choice(available_ecus))))
              for _ in range(4)],
        '2xB': [('B',
              available_ecus.pop(available_ecus.index(random.choice(available_ecus))))
              for _ in range(2)],
        'C+A': [('C',
              available_ecus.pop(available_ecus.index(random.choice(available_ecus)))),
          ('A',
                  available_ecus.pop(available_ecus.index(random.choice(available_ecus)))),],
        '2xA+B': [('A',
              available_ecus.pop(available_ecus.index(random.choice(available_ecus)))) ,
          ('A',
                  available_ecus.pop(available_ecus.index(random.choice(available_ecus)))),
          ('B',
                  available_ecus.pop(available_ecus.index(random.choice(available_ecus)))),],
        'D': [('D',
              available_ecus.pop(available_ecus.index(random.choice(available_ecus))))]
    }

    return scheme_mappings.get(scheme, [])
```


D. Mohammad Zaheri, *Safety-Oriented Optimization of Automotive E/E Architectures*, Wissenschaftliche Reihe Fahrzeugtechnik Universität Stuttgart, https://doi.org/10.1007/978-3-658-53045-7

Kolibri 360 GmbH
Ferdinand-Jühlke-Straße 7
99095 Erfurt, Deutschland
produktsicherheit@kolibri360.de